Professional Encounters in TESOL

Palgrave Studies in Professional and Organizational Discourse

Titles include:

Sue Garton and Keith Richards (*editors*)
PROFESSIONAL ENCOUNTERS IN TESOL
Discourses of Teachers in Teaching

Cecilia E. Ford
WOMEN SPEAKING UP
Getting and Using Turns in Workplace Meetings

Rick Iedema (*editor*)
THE DISCOURSE OF HOSPITAL COMMUNICATION
Tracing Complexities in Contemporary Health Care Organizations

Louise Mullany
GENDERED DISCOURSE IN THE PROFESSIONAL WORKPLACE

Keith Richards
LANGUAGE AND PROFESSIONAL IDENTITY

H. E. Sales
PROFESSIONAL COMMUNICATION IN ENGINEERING

Forthcoming titles include:

Edward Johnson & Mark Garner
OPERATIONAL COMMUNICATION

Palgrave Studies in Professional and Organizational Discourse
Series Standing Order ISBN 978–0–230–50648–0
(*outside North America only*)

You can receive future titles in this series as they are published by placing a standing order. Please contact your bookseller or, in case of difficulty, write to us at the address below with your name and address, the title of the series and the ISBN quoted above.

Customer Services Department, Macmillan Distribution Ltd, Houndmills, Basingstoke, Hampshire RG21 6XS, England.

Professional Encounters in TESOL

Discourses of Teachers in Teaching

Edited by

Sue Garton
Aston University

and

Keith Richards
University of Warwick

First published in 2008 by
PALGRAVE MACMILLAN
Houndmills, Basingstoke, Hampshire RG21 6XS and
175 Fifth Avenue, New York, N.Y. 10010
Companies and representatives throughout the world.

PALGRAVE MACMILLAN is the global academic imprint of the Palgrave
Macmillan division of St. Martin's Press, LLC and of Palgrave Macmillan Ltd.
Macmillan® is a registered trademark in the United States, United Kingdom
and other countries. Palgrave is a registered trademark in the European
Union and other countries.

ISBN-13: 978–0–230–55351–4 hardback
ISBN-10: 0–230–55351–6 hardback

This book is printed on paper suitable for recycling and made from fully
managed and sustained forest sources. Logging, pulping and manufacturing
processes are expected to conform to the environmental regulations of
the country of origin.

A catalogue record for this book is available from the British Library.

Library of Congress Cataloging-in-Publication Data

Professional encounters in TESOL : discourses of teachers in teaching /
edited by Sue Garton and Keith Richards.
    p.cm.—(Palgrave studies in professional and organizational discourse)
Includes bibliographical references and index.
ISBN 0–230–55351–6 (alk. paper)
    1. English language – Study and teaching – Foreign speakers. I. Garton,
Sue, 1961– II. Richards, Keith, 1952–

PE1128.A2P73 2008
428.2′4—dc22   2008014516

10 9 8 7 6 5 4 3 2 1
17 16 15 14 13 12 11 10 09 08

Printed and bound in Great Britain by
CPI Antony Rowe, Chippenham and Eastbourne

Contents

List of Tables and Figures

Tables

Figures

Notes on Contributors

Fiona Copland has worked as an English language teacher and teacher trainer in Nigeria, Hong Kong, Japan and England. She has been course director for a number of programmes including the MA Education (TEFL) at the University of Birmingham and CELTA and DELTA courses at the British Council, Tokyo, and at Brasshouse Language Centre, Birmingham. She is currently Programmes Director for the suite of MSc TESOL courses at Aston University.

Julian Edge is a senior lecturer in education at the University of Manchester, where he teaches on their master's courses in TESOL and supervises doctoral research in the area of teacher education. His overseas TESOL experience was gained in Europe, the Middle East and South East Asia. His 2002 book, *Continuing Cooperative Development*, is the publication he would most like people to read.

Sue Garton is Academic Director of the Centre for English Language and Communication at Aston University. She teaches on master's programmes in TESOL and on courses in English for Academic Purposes. Her main research interests are in language teacher education, classroom interaction and teaching EAP.

Amanda Howard has worked as an English language teacher, teacher educator and lecturer in the Middle East and United Kingdom, and has been involved in the inception and development of Young Learner English language teacher education courses at degree and masters level. She currently works on a freelance basis for British universities both in the United Kingdom and abroad, continuing her interest in teacher education and development.

Kuchah Kuchah has been a teacher trainer in Cameroon for over ten years. Now he works in the Ministry of Basic Education as National Pedagogic Inspector for bilingual education. He also serves as Secretary General and Chief Convenor of the Cameroon English Language and Literature Teachers' Association and as Events Coordinator for the Young Learners' Special Interest Group of the IATEFL. He holds an MA from Warwick University.

Fotini Vassiliki Kuloheri holds a BA in English Language and Literature (University of Athens), an MA in Applied Linguistics (University of

Essex) and the RSA Diploma with a Distinction. She is an experienced EFL teacher, materials writer, teacher educator and examiner for the Greek National Certificate of EFL. She is currently teaching at Greek primary state schools and is a doctorate student at the University of Warwick.

Nur Kurtoglu Hooton has 21 years' experience in the field of teaching English. She has been working at Aston University since 1994, managing, developing, and teaching on a range of programmes that have included academic English, communication skills, and teacher training courses both on-campus and distance learning. More recently she has also been working as a tutor on the MSc in TESOL and the MA in TESOL Studies programmes.

Wei Liu is currently studying on the MA TESOL and Translation Studies at Aston University. He has language teaching, learning and translating experience in Australia, China and the United Kingdom, across a range of fields including higher education, marketing, sports and foreign trade. He teaches both English and Mandarin on a voluntary basis.

Steve Mann is Associate Professor at the Centre for English Language Teaching Education (CELTE) at the University of Warwick. He has taught on a range of English language programmes and teacher education programmes in a range of institutions and companies in the United Kingdom, Europe, Hong Kong and Japan. He is particularly interested in supporting teaching development for TEFL teachers.

Muna Morris-Adams has extensive experience of teaching modern languages, English and self-defence in the United Kingdom. A graduate of Aston's MSc in TESOL, she is currently working as a tutor on the same programme, and on other postgraduate courses for teachers, both on-campus and distance learning. Her research interests include methodology, classroom interaction and intercultural communication.

Phil Quirke is Director of the Madinat Zayed College, Higher Colleges of Technology in the United Arab Emirate. He has been in ELT for over 20 years and has published on areas as diverse as the importance of face in the language classroom, action research, appraisal, the impact of laptop technology on the ELT classroom and the use of journals.

Keith Richards is an Associate Professor at Warwick University. His main research interests lie in the area of professional interaction and his recent publications include *Qualitative Inquiry in TESOL, Applying*

Conversation Analysis (edited with Paul Seedhouse), and *Language and Professional Identity*.

Paul Seedhouse is Professor of Educational and Applied Linguistics in the School of Education, Communication and Language Sciences, Newcastle University, UK. His monograph *The Interactional Architecture of the Language Classroom: A CA Perspective* was published by Blackwell in 2004. He also co-edited the collections *Applying Conversation Analysis* (Palgrave Macmillan 2005) and *Language Learning and Teaching as Social Interaction* (Palgrave Macmillan 2007).

Jerry Talandis Jr. is an English teacher at the Toyama College of Foreign Languages, in Toyama City, Japan. He has taught in Japan since 1993 to a wide range of levels and students, from children to senior citizens. He is particularly interested in educational technology, especially supporting and researching online TEFL teacher development.

Maneerat Tarnpichprasert is a lecturer at the Faculty of Education, Chulalongkorn University, Thailand. She is currently studying for a PhD in ELT and Applied Linguistics at the University of Warwick. Her research interests are bilingual education and teachers' professional identity.

Sue Wharton is Associate Professor at the Centre for English Language Teacher Education, Warwick University, UK. She teaches on BA and MA courses and supervises research students. She is particularly interested in the discourse of TESOL research and in discoursal representations of the family.

Transcription Conventions

The following transcription conventions have been used throughout the book, although the level of detail in the transcriptions varies from author to author.

[	point of overlap onset
]	point of overlap termination
=	a) turn continues below, at the next identical symbol b) if inserted at the end of one speaker's turn and at the beginning of the next speaker's adjacent turn, indicates that there is no gap between the two turns
(3.2)	length of a timed pause
(.)	very short untimed pause
<u>word</u>	speaker emphasis
e:r the:::	lengthening of the preceding sound
-	abrupt cut-off
?	rising intonation, not necessarily a question
!	animated or emphatic tone
,	low-rising intonation, suggesting continuation
.	falling (final) intonation
CAPITALS	especially loud sounds relative to surrounding talk
↑↓	marked shifts into higher or lower pitch in the utterance following the arrow
° °	utterances between single degree signs are noticeably quieter than surrounding talk
°° °°	utterances between double degree signs are considerably quieter than surrounding talk
> <	utterances between inward-facing arrows are produced more quickly than surrounding talk

< >	utterance between outward-facing arrows are produced more slowly and deliberately than surrounding talk
( ) or (xxx)	stretch of unclear or unintelligible speech.
(guess)	transcriber doubt about a word
.hh	speaker in-breath
hh	speaker out-breath
→	arrows in the left margin pick out features of interest
((T shows picture))	non-verbal actions or author's comments
ja *yes*	translations into English are italicized and located on the line below the original utterance
[gibee]	in the case of inaccurate pronunciation of an English word, an approximation of the sound is given in square brackets
[æ]	phonetic transcriptions

Introduction

Sue Garton and Keith Richards

Ten years ago Johnston asked provocatively, 'Do EFL teachers have careers?' The fact that the jury is still out would not surprise him: 'the question', he concluded, 'will have to go unanswered.' (1997: 707). For him, though, it was the wrong question to begin with. Teachers' lives, he argued (Ibid.: 708), 'are lived in complex contexts in which personal, educational, political and socioeconomic discourses all influence the way the life is told'.

The study of such lives, which has a long and respectable pedigree in mainstream educational research, has received only patchy attention in TESOL, though we are now at a point where those whose teaching world was transformed by the communicative revolution are approaching the threshold of retirement. It therefore seems a natural time to examine the sorts of experiences that help shape a working life in our chosen profession, to share understanding and insights among generations, and not least to celebrate what is rich and interesting about our lived world. In that realisation lies the genesis of this book.

Although we will use terms such as 'profession' and 'career', the question of whether TESOL is indeed a profession or even a career is not one that we will consider. In a global context such as ours, the meaning of the question itself varies so much from place to place that the pursuit of any general answer would be fruitless. However, our growing sensitivity to local educational ecologies (Holliday 1994) and rejection of Copernican assumptions about English language teaching (see, for example, Canagarajah 2005; Holliday 2005; Pennycook 1994) does not entail an outright rejection of common experiences in the profession. Prolonged engagement with the many professional worlds of TESOL involves a process of growth and development which fosters the refinement of craft skills, professional knowledge and experiential resources, features of which will be common to most if not all teachers.

Such skills and knowledge are developed through encounters with the professional world and instantiated through talk. Our focus in this book is on the discourses that English language (EL) teachers encounter throughout their careers. We look at the reflexive nexus of discourse, practice and development, at how professional development takes place through professional conversations (Crookes 1997: 68) and is in part defined by these. By analysing the developing discourses of teachers in teaching and training, we aim to shed light on what EL teachers do and why they do it.

The concept of career stages serves as a useful organisational device, but these are interpreted very broadly in terms that will be recognisable to those involved in teaching. Our real interest lies in the discursive encounters of teachers at different points in their careers and what can be learnt from these. TESOL is well served in terms of available training and development courses, both academic and practical, but little attempt seems to have been made to stand back and look at the career trajectory as a whole. Perhaps this is merely because up to now such an attempt would have been premature, but it nevertheless encourages a view of TESOL as a profession temporally as well as geographically compartmentalised.

Our aim, then, is to explore the changes that take place in a teacher's life through some of its constituent discourses. At the same time, we hope to reveal something of the subtle changes in these discourses that take place over the career cycle.

Organisation of the book

A full teaching lifetime will see many encounters, some of them wearingly predictable (Doyle 1986, for example, refers to the finding that an average elementary school teacher will publicly praise or blame pupils 16,000 times a year) and others utterly transformative. In this book we have imagined such a lifetime and have tried to trace the teaching world as it might seem from the perspective of a teacher moving through the different stages of the career trajectory.

In order to do this we have taken as our point of orientation Huberman's classic description of the teacher career cycle (Figure 0.1). This is by his own admission an imperfect representation of a complex and varied journey and it is by no means the only description available (for similar work, see Fessler and Christiansen 1992; Sikes, Measor, and Woods 1995), but it nevertheless provides one of the most useful general routemaps available to us.

The basic model will probably be familiar to most teachers on the basis of their own professional experiences and observations, and although it is impossible to reflect the full range of encounters and discoveries that characterise each stage, the chapters that make up this volume address what we believe to be a representative range of issues. We have omitted

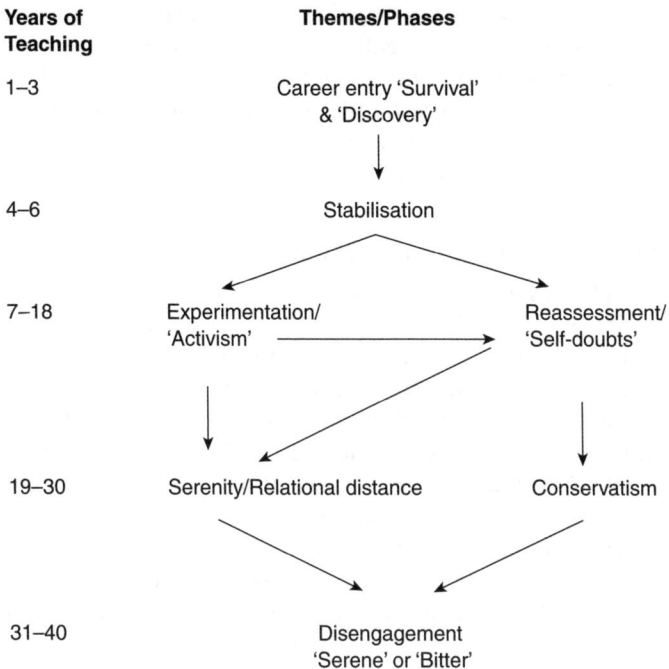

Years of Teaching	Themes/Phases
1–3	Career entry 'Survival' & 'Discovery'
4–6	Stabilisation
7–18	Experimentation/'Activism' — Reassessment/'Self-doubts'
19–30	Serenity/Relational distance — Conservatism
31–40	Disengagement 'Serene' or 'Bitter'

Figure 0.1 Huberman's themes of the teacher career cycle (1992: 127)

only the final stage of Huberman's model because professional engagement is a core dynamic of the collection and the issues raised by the process of disengagement fall outside this. Similarly, in Part III we have focused on seeking new horizons rather than reaching a professional plateau, which has been recognised as a distinct stage by some researchers (see Day, Stobart et al. 1996: 174). In taking this view we might be criticised for excessive optimism, but the orientation of all involved in this project is to the possibility of productive engagement.

This leaves four stages, made up as follows:

Part I *Starting Out* (survival and discovery)

This section focuses on previous learning experiences, initial teacher training and the classroom experience of the beginning teacher. One of the key elements in any introductory training programme is teaching practice and the feedback on this. The first two chapters focus on this vital area, Copland from the perspective of the trainee and Kurtuglou Hooton on the nature of the feedback itself. Taken together, these two chapters provide important insights that will help those involved maximise the impact of this essential encounter. The other important form of learning, from classroom experience, is addressed by Seedhouse, who uses classroom evidence to highlight some of the important differences between the classroom interaction of novice teachers and that of their more experienced colleagues.

Part II *Becoming Experienced* (stabilisation)

The emphasis here is on the sort of experiences and challenges that shape the more established teacher's development. Garton's chapter links the classroom discourse that was the focus of Seedhouse's chapter with teacher beliefs, aiming to raise awareness of the extent to which particular belief systems impact on classroom behaviour. Howard's contribution tackles an aspect of teaching that looms large in teachers' professional world but has been strangely neglected by researchers: the observed lesson. It explores what happens when experienced teachers are observed in the classroom for the purposes of appraisal and suggests ways of making the most of this challenging situation. Finally, Morris Adams shifts the focus outside the classroom and explores how teachers can productively engage with the linguistic experiences of their students outside the classroom environment.

Part III *New Horizons* (experimentation/reassessment)

Many teachers reach a point in their careers where they feel the need to explore new aspects of their professional world. One way of doing this is through study at Master's level, and this has been the focus of interesting research, some of it from the perspective of discourse development (see, e.g., Cutting 2000; Freeman 1991, 1992). However, there are many other dimensions of exploration which seem to have been neglected. Quirke's chapter is set in the context of a Master's programme but looks at the potential of the web as a medium for sharing professional knowledge and understanding. Mann's chapter stays on the theme of sharing but looks at it from the perspective of developing ideas with fellow teachers through professional talk. Richards's contribution focuses on a situation where a group of teachers were given the rare opportunity to start a school from scratch. It describes the process of establishing the school, relating this to the teachers' aims and the eventual development of distinctive ways of interacting.

Part IV *Passing on the Knowledge*
(serenity/relational distance)

There are many ways in which teachers might pass on knowledge; in fact, it is something that goes on in small but not insignificant ways in staffrooms around the world. Those who wish to communicate their ideas and understandings to a wider community, however, must look beyond this to the potential of teacher training or publication. Kuchah Kuchah presents the challenges of professional development in sub-Saharan Africa as he explores the teaching/learning culture of power relationships and its discourses during his journey from class-room teacher to one of the youngest school inspectors in Cameroon. Wharton addresses the process of writing for publication, showing what this involves, professionally and discoursally, for teachers who wish to develop their ideas into published works. Finally, in a fitting conclusion to the collection, Edge reflects on 40 years in TESOL, exploring dimensions of evaluative discourse and their developmental potential.

The focus of this book is on practising English language teachers and their concerns at different stages in their professional lives. The selection of topics covered has been informed by a desire to offer insights and illustrations that will allow teachers to rethink aspects of their experiences and opportunities at particular stages in their development, so although the chapters are research-based they are orientated to the practical needs of teachers. Because the voices of teachers are heard only indirectly, and usually within an analytical framework which may act as a selectional and interpretive filter, we felt it was also important to hear the voice of the teacher directly.

Each of the four parts is therefore not only preceded by an introduction setting the chapters in their conceptual context, but also followed by a reflection from a practising teacher at that stage in his or her career. The teachers, chosen to represent not only different career stages but also different cultural and experiential settings, were invited to read and comment from their own perspectives on the chapters in their section. They were given the following points as a guide to follow, if they wished:

- Where am I now? What am I doing? Where have I come from and where am I going?
- What matters to me now? What are the issues that concern me?
- How far do these chapters address these issues? What's most important about them?

- What's missing? What lessons can I pass on, if any?
- If I had one message to pass on to someone approaching this phase in their career cycle, what would it be?

To ensure that the teachers' voices are authentic, no editorial control was exercised over their contributions, which have been left exactly as they were written, save minor corrections to typing errors. Considered along with the issues raised in the introductory section, the resulting observations should serve as a useful starting point for reflection and discussion. Inevitably, much of this will be from the perspective of a particular career stage, but the presence of views from different stages also offers the possibility of broader contextualisation in terms of the career trajectory as a whole and the unfolding relationship between knowledge, discourse and action.

The career trajectory

Globally, English language teaching is worth billions of pounds and it depends – even in a world saturated comprehensively with communications technology – on the daily efforts of the teachers whose knowledge and experience, gathered through a lifetime of professional development, contribute immeasurably to its success. Yet, despite the huge literature generated to serve the interests of the field, until recently the teacher remained an almost invisible figure, an assumed constant in a history marked by dramatic changes of approach and orientation. Huberman perfectly captures the resulting distortion of perspective:

> When one overlooks people's lives to focus on events...or on the institutional theatres of those events, one is taking the actors out of the play and assuming that the scenery is animate enough to carry the plot and account for the denoument.
>
> (Huberman 1988: 120)

The consequences of this have perhaps not been fully appreciated. Books designed for teachers, such as those focusing on methods, materials or professional development, have traditionally adopted what might be described as an *all-life approach*, which ignores the changing perspectives that characterise a teaching life. The new teacher struggling to master the essential skills of day-to-day planning, preparing, teaching and staffroom socialising will not respond to new ideas or fresh initiatives in the same way as an experienced colleague. Over time, perspectives shift as contexts and colleagues change, skills are refined, knowledge develops and views mature. Although it is clearly neither feasible nor necessary (since teachers are more than capable of assessing the relevance or otherwise of a piece of advice in terms of their own professional circumstances) to organise books for teachers in terms of career stages, there is nevertheless a need for greater recognition of the importance of this aspect. This collection offers an approach that, by highlighting key encounters at each stage in the career trajectory, maximises the potential for knowledge growth and development at that stage.

It has been argued that TESOL is a 'permeable' profession (Maley 1992) and there is certainly some truth in this claim, but it is far less true now than it once was. Hayes, who has perhaps contributed more than anyone to research into ESOL teacher careers (see, e.g., Hayes 1996, 1997), challenged this view directly in a recent paper based on the life stories of three Sri Lankan teachers (2005). For these teachers the profession is seen as something which, once entered, follows a distinct career path through to retirement. Interestingly, while not suggesting that career progression is unproblematic, Hayes is able to capture through these teachers' voices a sense of remarkably uncomplaining commitment, sustained enthusiasm and undiminished energy late into the career in the face of sometimes extreme threats, including physical violence and political upheaval.

Some also claim that the concept of a career change model is unduly restrictive. While still broadly accepted by many researchers, it has been criticised as failing to reflect the complex and idiosyncratic nature of individual career trajectories (e.g. Cheung 2005). This seems to us to misunderstand the purpose of such models, which are intended not as pigeon-holes but as thematic characterisations in terms of which individual experiences might be interpreted. Work in this area must be distinguished from research into life stories, which needs to be much more richly contextualised and demands a different interpretive stance. Simon-Maeda's study (2004) of the construction of professional identities of female EFL teachers in Japan, for example, probes deeply into personal experiences in order to shed light on the conflictual dynamics of personal and professional engagement with wider institutional and sociocultural forces. The issues of marginalisation and discrimination that it raises have not only general relevance but personal potency.

While the career trajectory highlights aspects of change, it is also important not to exaggerate differences among teachers, however varied their backgrounds and experience. In an illuminating and fascinating investigation into the distinctive characteristics of foreign language teachers, Borg (2006) reveals some interesting features. One of the strengths of his study lies in the range of its sample, which covered five distinct groups representing over 200 respondents: teachers on a post-graduate TESOL course, language teacher conference delegates, university teachers of other subjects, Hungarian pre-service teachers of English, and Slovene undergraduates in English. The picture that emerges is of a dynamic subject, unique in scope and complexity and with a more diverse teaching methodology than other subjects. Particularly interesting is the fact that language teachers, though they are subject to greater commercial pressures than other subject teachers, place particular value on creativity, flexibility and enthusiasm.

The model used here, then, is a convenient organising device, a useful heuristic rather than a binding interpretive frame, and our decision to include the voice of the individual teacher as discussant recognises the importance of the link between the interpreted world and the lived world, where the qualities identified by Borg are particularly salient. It is almost certainly true that no teacher reading this book will see in it an exact reflection of their own experiences, but neither should they find anything here alien to those experiences, nor anything in the treatment of teacher knowledge that underestimates the complexities of professional engagement.

Teacher knowledge

Central to this book is the consideration not only of what teachers do but also of why they do it and how this is revealed through their discourses. Perhaps surprisingly, the realisation that 'in order to better understand language teaching we need to know more about language teachers' (Freeman and Richards 1996: 1) is relatively recent (Richards 1994). However, this realisation has led us away from a product-process view of teaching, where teachers are seen to have a body of content knowledge to transmit to students and a set of methodological principles with which to do so (Freeman 2002: 4) to the recognition of a complex interrelationship between what teachers know, how they learn and their classroom practice.

According to Tsui (2003: 7) teacher knowledge includes such aspects as knowledge of subject matter, learners, the curriculum, the context in which they work and pedagogy, but she follows Cumming in recognising that the kinds of knowledge teachers draw on in their work 'appear to exist largely in very personalised terms, based on unique experiences, individual conceptions, and their interaction with local contexts' (1989: 46–7, cited in Burns 1992: 57–8).

Such knowledge is context dependent and best seen as developing in a reflexive relationship with the environments in which teachers work. The importance of interaction in the construction of teacher knowledge and understanding has been underlined by Tsui, who notes of the teaching context:

> This includes their interactions with people in their contexts of work, where they constantly construct and reconstruct their understandings of their work as teachers.
>
> (Tsui 2003: 64)

Since it is these contexts of work, the interactions that take place there and the consequent construction/reconstruction of teachers' work that are the focus of this book, we now move on to consider the development of teacher knowledge in some of these contexts and in key areas of their professional lives.

Starting out

At the start of their teaching careers, it may be thought that the most important influence on a new teacher is pre-service teacher education. Yet research has shown that this is often not the case and beginning teachers are often more strongly influenced by their own experiences as learners than by the input from their teacher education programme (Tatto 1998). In particular, pre-service teachers already have strong ideas about teaching as a result of what Lortie (1975) calls 'the apprenticeship of observation', the years spent as pupils observing what it means to be a teacher.

Bailey and colleagues (1996), for example, examined the influence of a teacher's own experience as a learner on how they teach. Their research revealed that ideas about 'good' or 'bad' teaching were strongly influenced by models that teachers were exposed to as learners. Moreover, even in cases where training did seem to have an effect, when faced with difficult situations in the classroom the teachers reverted to teaching in the way they themselves had been taught and not in a way conducive to the learning environment that the training programme wished to encourage. The authors conclude that the 'apprenticeship of observation' is very important in determining how trainees teach and even where new approaches have been learned in teacher training, these do not seem to be available to trainees during difficult moments in the classroom.

An interesting question in initial teacher education concerns the extent to which trainees may pay lip service to what they are taught during their training because they need to pass the course, while their underlying ideas remain basically unchanged. Guttiérez Almarza (1996), for example, notes that while there was evidence that the teacher education programme did play a role in her trainees' practice, the way they talked about this practice was deeply rooted in their pre-training assumptions and experience.

It would seem, therefore, that it is very difficult for teachers to change deep-rooted and long-standing ideas they have about teaching and learning. Johnson (1994) found that previous formal learning experience was the most powerful influence on teachers' images of themselves as teachers, on their teaching and on their perceptions of their own practices. Even though they were critical of their teacher-centred practices, the teachers in Johnson's study found that they could not change. Any conceptions of alternative practices were based on informal experiences as language learners, but they found that they were unable to translate these experiences into classroom practice.

The above discussion indicates two important aspects of the discursive encounters involved in *starting out*. First of all, detailed analysis is needed of the discourses of initial teacher education in order to investigate how trainers and trainees construct teacher knowledge, in what areas and how the apprenticeship of observation may act as a filter through which new knowledge is interpreted. Secondly, it is necessary to look at the discourses of post-course experience for evidence of more deep-rooted change that goes beyond conforming to the expectations of the course. This is an area that has been very much neglected in TESOL to date.

Becoming experienced

As teachers move through their careers, the experiences they gain in the context of their teaching will affect their knowledge and practices. In order to better understand how knowledge develops and evolves as teachers become more experienced, we need to look not only at what teachers do in their day-to-day professional lives but also why they do it. The way in which teachers talk about their experiences is fundamental to understanding how a teacher's knowledge influences what happens in their context of work, and is in turn influenced by what happens there.

Burns (1992) identified three interconnected contextual levels that play a role in the development of the knowledge of experienced teachers. Firstly, there is the institutional level which concerns the 'ways in which particular organisational ideologies or philosophies were interpreted by the teachers' (Burns 1992: 158). Secondly there is the classroom level which concerns teacher knowledge and beliefs about language, language learning and the learners themselves. Finally, there is the instructional level which concerns actual lesson planning and decisions made in terms of tasks, materials and the teacher's role.

Although models for understanding teachers' classroom decision-making have been proposed (e.g. Woods 1996), Burns's contextual levels show that the world of professional experience is much wider. The three levels are presented as distinct, but they are clearly interconnected. Institutional discourses may also impinge on the classroom itself and this is the case of the discourse of the evaluation and assessment of teachers' work, for example. Moreover, what teachers do in their classrooms will have an effect outside as their learners interact in society. This is another neglected area of research in our field which is explored in this collection.

New horizons

As teachers become increasingly experienced, many feel the need for further professional development. Some may opt for one of the more advanced formal teacher education programmes that are available; others may decide to stay in their classrooms while looking for alternative forms of professional development; still others may decide to move out of their classrooms altogether and take on a different role within the profession. Some may do all three.

For those who decide to undertake a more formal course, there is a question as to the extent to which teacher education influences teacher knowledge beyond just acquiring the discourse. Freeman (1991), for example, looked at how experienced foreign language teachers' conceptions of their practice evolved as a result of the development of a shared professional discourse that emerged during a teacher education programme. He found that as a result of being able to better articulate their thinking using the shared discourse, the teachers gained greater control over their classroom practice. However, the question remains open as to whether the shared discourse simply enabled the teachers to 'make the tacit explicit' (to use Freeman's words), that is, the conceptions they articulated already existed before the training programme, or whether it was participation in the training programme that led to the formation of these conceptions. Freeman concludes that both processes were actually 'taking place simultaneously and interactively' so that there seems to be a dialectical relationship 'in which familiar and tacit knowledge interacts with – and is shaped by – newly explicit understandings' (1991: 453).

Today the development of shared discourses is no longer limited to formal teacher education. Advances in information technology mean that experienced teachers from all over the world can now share and support each other in their professional development in ways such as those described by Quirke in this volume. It seems likely that this will give rise to new discourses and shape teacher knowledge in new ways as teachers are able to interact more easily with their peers rather than have to seek knowledge in a formal context. Although research into this area is still very much in its infancy, the importance of analysing these new discourses is clear.

Passing on the knowledge

As we enter the final stage in the career trajectory outlined in this book, there is a sense in which we have come full circle. The discourses of teachers who *pass on the knowledge* will shape the beliefs and ideas of those who are just *starting out*. However, there is another sense in which the cycle never closes as the knowledge base (Verloop, Van Driel and Meijer 2001) of even the most experienced teachers continues to be shaped by such encounters, both formal and informal, spoken and written.

Those who have written the chapters in this book have passed through all the stages in the career trajectory outlined here. The very act of writing a chapter is to participate in the discourse of passing on the knowledge. But it is equally important that the voices of those who are at different stages of the career cycle are also heard as only they can tell us how relevant the professional encounters that we have described are to them. This is why we have invited practising teachers to offer their reflections and share their experiences in terms of the particular stage of the career cycle that is relevant to them. Those who are reading or working with this book thus have the opportunity to respond not only to the meta-perspective of those who have contributed chapters, but also to the insider views of those who are experiencing the encounters in the here and now.

The world of professional experience in TESOL is vast and in this collection we seek to make connections with a variety of contexts in which decision-making and knowledge construction are realised through discursive encounters. In the final analysis, it is what teachers do in the classroom that counts, but Heraclitus's observation that it is impossible to step into the same river twice might easily be extended to the classroom. The development of the teaching experience over 40 or more years, from initial encounters to mature engagement, is subject to a myriad of influences that make the career trajectory – and the knowledges associated with it – a fascinating object of study.

Conclusion

Although this collection can be read as it stands, working through the career trajectory in its natural temporal sequence, at least one other approach suggests itself. Readers can begin with the section related to their current career stage, reflecting on its relevance or otherwise to their present situation and comparing their views with those of the teacher-reviewer included here. Previous and subsequent career stages can then be explored, so that by relating this 'now' to both perceived past and possible future, readers can explore their current situation and plans from fresh perspectives.

There is a sense in which we build our careers retrospectively, interpreting all that has gone before in terms of a coherent narrative developing to the point we have now reached, producing an ordered progression of tidy events and clear decisions. We hope that this collection will offer a more nuanced perspective and that its contribution will lie not so much in any general claims it might make as in the power of the particular to resonate and to stimulate change:

> The powerful moment, the moving insight (though sometimes just from one person or even a handful) is sometimes enough to create dynamic involvement in those who have access to it.
>
> (Schubert 1990: 100)

References

Bailey, K. C., Bergthold, B., Braunstein, B., Jagodzinski Fleischman, N., Holbrook, M. P., Tuman, J., Waissbluth, X. and Zambo, L. J. 1996. 'The language learner's autobiography: Examining the "apprenticeship of observation"'. In D. Freeman and J. C. Richards (eds), *Teacher Learning in Language Teaching*. Cambridge: Cambridge University Press, pp. 11–29.

Borg, S. 2006. 'The distinctive characteristics of foreign language teachers'. *Language Teaching Research*, 10(1): 3–31.

Burns, A. 1992. 'Teacher beliefs and their influence on classroom practice. *Prospect*, 7(3): 56–66.

Canagarajah, S. (ed.) 2005. Reclaiming the Local in Language Policy and Practice. Mahwah, NJ: Lawrence Erlbaum.

Cheung, E. 2005. 'Hong Kong secondary schoolteachers' understanding of their careers'. *Teachers and Teaching: Theory and Practice*, 11(2): 127–49.

Crookes, G. 1997. 'What influences what and how second and foreign language teachers teach?' *The Modern Language Journal*, 81(1): 67–79.

Cumming, A. 1989. 'Student teachers' conceptions of curriculum: Towards an understanding of language teacher development', *TESL Canada Journal*, 7(1): 33–51.

Cutting, J. 2000. *Analysing the Language of Discourse Communities*. Amsterdam: Elsevier.

Day, C., Stobart, G., Simmons, P. and Kington, A. 2006. 'Variations in the work and lives of teachers: Relative and relational effectiveness'. *Teachers and Teaching: Theory and Practice*, 12(2): 169–92.

Doyle, W. 1986. 'Classroom organization and management'. In M. C. Wittrock (ed.), *A Handbook of Research on Teaching* (3rd edn). New York: Macmillan, pp. 392–431.

Fessler, R. and Christensen, J. 1992. *The Teacher Career Cycle: Understanding and Guiding the Professional Development of Teachers*. Boston, MA: Allyn & Bacon.

Freeman, D. 1991. '"To make the tacit explicit": Teacher education, emerging discourse, and conceptions of teaching'. *Teaching and Teacher Education*, 7(5/6): 439–54.

Freeman, D. 1992. 'Language teacher education, emerging discourse, and change in classroom practice. In J. Flowerdew, M. Brock and S. Hsia (eds), *Perspectives in Second Language Teacher Education*. Hong Kong: City Polytechnic of Hong Kong, pp. 1–21

Freeman, D. 2002. 'The hidden side of the work: Teacher knowledge and learning to teach', *Language Learning*, 35, 1–13.

Freeman, D. and J. C. Richards (eds), 1996. *Teacher Learning in Language Teaching*. Cambridge: Cambridge University Press.

Gutiérrez Almarza, G. 1996. 'Student foreign language teacher's knowledge growth'. In D. Freeman and J. C. Richards (eds), pp. 50–78.

Hayes, D. 1996. 'Prioritizing "voice" over "vision": Reaffirming the centrality of the teacher in ESOL research'. *System*, 24(2): 173–86.

Hayes, D. 1997. 'Articulating the context: INSET and teachers' lives'. In D. Hayes (ed.), *In-service Teacher Development: International Perspectives*. Hemel Hempstead: Prentice Hall, pp. 74–85.

Hayes, D. 2005. 'Exploring the lives of non-native speaking English educators in Sri Lanka'. *Teachers and Teaching: Theory and Practice*, 11(2): 169–94.

Holliday, A. 1994. *Appropriate Methodology and Social Context*. Cambridge: Cambridge University Press.

Holliday, A. 2005. *The Struggle to Teach English as an International Language*. Oxford: Oxford University Press.

Huberman, M. 1988. 'Teacher careers and school improvement'. *Journal of Curriculum Studies*, 20(2): 119–32.

Huberman, M. 1992. 'Teacher development and instructional mastery'. In A. Hargreaves and M. G. Fullan (eds), *Understanding Teacher Development*. London: Cassell, pp. 122–42.

Johnson, K. E. 1994. 'The emerging beliefs and instructional practices of preservice English as a Second Language teachers'. *Teaching and Teacher Education*, 10(4): 439–52.

Johnston, B. 1997. 'Do EFL teachers have careers?' *TESOL Quarterly*, 31(4): 681–712.

Lortie, D. 1975. *Schoolteacher: A Sociological Study.* Chicago: University of Chicago Press.

Maley, A. 1992. 'An open letter to "the profession"'. *ELT Journal*, 46(1): 96–9.

Pennycook, A. 1994. The Cultural Politics of English as an International Language. *London: Longman.*

Richards, K. 1994. 'From guessing what teachers think to finding out what teachers know: The need for a research agenda'. *TESOL Quarterly*, 28(2): 401–4.

Schubert, W. H. 1990. 'Acknowledging teachers' experiential knowledge: Reports from the teacher lore project'. *Kappa Delta Pi Record*, Summer, 99–100.

Sikes, P. J., Measor, L. and Woods, P. 1985. *Teacher Careers: Crises and Continuities.* London: Falmer Press.

Simon-Maeda, A. 2004. 'The complex construction of professional identities: Female EFL educators in Japan speak out'. *TESOL Quarterly*, 38(3): 405–36.

Tatto M. T. 1998. 'The influence of teacher education on teachers' beliefs about purposes of education, roles and practice'. *Journal of Teacher Education*, 49(1): 66–77.

Part I
Starting Out

Paula

With a good first degree in English Language and Literature, Paula has decided to become an English teacher in her country, where English is a foreign language. She has many years' apprenticeship of observation, observing language classrooms from the point of view of the learner, but realises that she needs some formal teacher training if she is to feel confident in her new profession. So, she signs up for an initial teacher training programme. Like so many teachers who are starting out, the element of her new programme that concerns her the most is the teaching practice component. What will be expected of her? Will she understand what she is supposed to do? What will happen if she does not do it 'right'?

Having just taught her first teaching practice lesson, Paula is reasonably satisfied that it went quite well, but that is only half the story. Now she must go to the post-observation group feedback session and hear what her tutor and her peers thought of her lesson. As she is not really sure what to expect, it is with some trepidation that she approaches her first feedback session.

Post-observation feedback is perhaps the single most important discourse that trainee teachers engage in and their success or failure may depend on their ability to deal with it effectively. Given its central role in initial teacher training courses and the concern it causes for those like Paula who are Starting Out, the first two chapters in this section both deal with this key speech event. The link between teacher preparation and the first year of teaching is a vital one (see, e.g., Liston, Whitcomb and Borko 2006) and the third chapter takes us into the classroom to look at what trainee teachers do compared to experienced teachers. In doing so, it also acts as a bridge between this part and the next.

In the first chapter in this section, Fiona Copland notes what an emotional experience an initial teacher training course, and especially the feedback event, can be. She looks at the feedback event as a genre and shows how trainers may have expectations of trainees that the latter are not aware of, that trainees may not be aware of because they are unfamiliar with the genre. By focusing on the talk of both trainees and trainers, Copland shows how important, but also how difficult it can be for trainees to understand the feedback event and to negotiate their way through what she calls its structures and conventions.

The significance of Copland's chapter is made clear as she illustrates how failure to come to grips with the trainer's expectations can have serious consequences for trainees as they may find themselves evaluated negatively. The chapter concludes by offering some important suggestions as to how trainees may be helped to understand the feedback event and what is expected of them.

Nur Kurtuglu Hooton's chapter is also concerned with the feedback event, but here the focus is on the actual feedback given and, in particular, on the effects that it has on trainees. Kurtuglu Hooton focuses on two different types of feedback, what she describes as corrective and confirmatory feedback, and the way they may not only bring about a growth in the knowledge of the student teachers, but also lead to change.

Most feedback in initial teacher training programmes tends to be corrective in nature, but Kurtgulu Hooton shows in particular that, while this type of feedback may ensure that the student teachers fulfil the expectations of the course, it is confirmatory feedback – praising teachers for what they did well – that is more likely to bring about deeper and more lasting change. The chapter suggests that, far from just paying lip-service to what they were taught, the student teachers who received confirmatory feedback underwent profound change, not only in their teaching, but also in the way they saw themselves.

Both these chapters may go some way to answering the question we posed in the introduction as to how initial teaching training can overcome the strong influence that trainees' initial learning experiences have on how they teach. By effectively training the trainees into the feedback event and making expectations clear, while at the same time shifting the focus of feedback more towards what trainees do well, trainers may be able to bring about more deep-rooted change.

In the final chapter in this section, Paul Seedhouse takes us into the classroom and addresses one of the main concerns that trainee teachers have about their classroom talk – that of creating a pedagogical focus through instruction-giving. Noting the frequent mismatch between what trainee teachers want students to do and what they actually do, Seedhouse uses Conversation Analysis to analyse the instruction-giving of both inexperienced and experienced teachers.

Seedhouse shows how a micro-analysis can reveal exactly how experienced teachers are able to establish and shift pedagogical focus, while inexperienced teachers may find it more challenging.

Part I closes with the reflections of Wei Liu on what it means to be Starting Out as a teacher whose apprenticeship of observation took place in Chinese classrooms, while his initial teacher training has taken place in Britain. Liu relates his own experiences closely to the situations described in the three chapters by giving a vivid description of his own process of starting out. Perhaps most significant is his focus on the importance for trainee teachers of 'being' as well as 'doing', which is clear not only in his comments on the chapters in this section, but also in his reflections on his own teaching.

References

Liston, D., Whitcomb, J. and Borko, H. 2006. 'Too little or too much: Teacher preparation and the first years of teaching'. *Journal of Teacher Education*, 57(4): 351–8.

1
Deconstructing the Discourse: Understanding the Feedback Event

Fiona Copland

Introduction

Joining an initial English language teacher training course can be exciting, confusing and traumatic all at the same time. It is exciting because it signifies a new start, perhaps a new career, and for many, a different approach to teaching and learning. It is confusing because there is a great deal of 'jargon' in English language teaching which trainees need to understand and begin to use. And it is traumatic because in most courses there is a teaching practice element in which trainees try out new skills and have their teaching assessed.

An element of such courses which can be exciting, confusing and traumatic simultaneously is the feedback event. This is usually held soon after teaching practice and is the opportunity for trainees and trainer to discuss the strengths and weaknesses of the lesson and, often, of the trainees. There is a great deal at stake here for both trainers and trainees: trainers must provide useful and accessible advice to trainees; trainees must demonstrate an ability to reflect and evaluate teaching and learning and take on board criticism and advice in order to improve their performance.

This chapter will deconstruct the discourse of the feedback event. Through an analysis of the feedback event as genre, it will outline its structures and conventions and suggest that trainees who are able to negotiate these are likely to be successful on the training programme. As part of the genre analysis, it will highlight trainers' expectations of the trainees and suggest ways in which trainees can fulfil these expectations. Finally, it will consider how trainees can be prepared to take part in feedback events in order to take the best advantage of the opportunities they afford.

Participants, setting and data

Throughout the chapter I will be illustrating the discussion with trainer and trainee talk. This talk was recorded during two initial teacher training courses in the post-observation feedback event at a language school in the Midlands. The feedback event was attended by the trainer and all the trainees (the term used here for the pre-service teachers) who had either taught or observed the lesson, a common practice on teacher training courses of this kind. Altogether, 16 trainees and four trainers participated in the research. Each feedback session usually lasted for between 45 minutes and an hour.

The research constitutes a linguistic ethnography (Rampton 2007). Linguistic ethnography is an emerging research space which brings together ethnographic and linguistic data collection and analysis research tools (Creese forthcoming). It highlights the importance of context in understanding talk-in-interaction and demands a high level of reflexivity from researchers in their description and analysis of data. In terms of this research, the data collection tools adopted were: fieldnotes from participant observation; trainer and trainee interviews; audio and video recordings of feedback and interviews with participants. The tools of analysis were content analysis of fieldnotes; linguistic and paralinguistic analysis of the recorded feedback drawing on the Conversation Analysis; and genre analysis. As stated above, this chapter will focus in particular on the feedback event as genre as this provides a useful gateway into learning how feedback is structured and the expectations such structuring places on participants. I am grateful to both the trainers and trainees who allowed me to observe and record the feedback sessions in which they took part and who took part in the interviews in their own time.

The feedback event as genre

The post-observation feedback session is a communicative event peculiar to teacher training and teacher evaluation. It is usually held soon after the teaching has taken place and is led by the trainer who has observed the lesson. One way of understanding it is to see it as a genre.

Genres have been described as follows:

> Conventionalised expectations that members of a social group or network use to shape and construe the communicative activity that they are engaged in. These expectations include a sense of the likely tasks on hand, the roles and relationships typically involved, the ways the activity can be organised, and the kinds of resources suited to carrying it out.
>
> (Rampton 2006: 128)

Through a process of observing and practising delivering feedback as they prepare to take on the training role, and then through acting and re-enacting the feedback event in the training role, trainers understand and perform the genre, for the most part effortlessly. However, trainees are rarely aware before joining the course what the feedback genre entails and must learn very quickly how to negotiate the framework it creates. Indeed, whether trainees pass or fail the course is in part determined by how successful trainees are at this negotiation.

The following sections uncover the 'conventionalised expectations' of the feedback event in terms of the tasks on hand, roles and relationships, organisation and resources. Data excerpts show how these expectations are realised by trainers and how trainees respond either successfully or unsuccessfully to the expectations placed on them.

Tasks on hand

Feedback is conducted for two main reasons: to provide an opportunity for participants to talk about what happened in the lesson and to provide trainees with feedback on the lesson. I will take talk about these two things to constitute the 'tasks on hand' of the feedback event.

In terms of 'what happened', anything connected with the teaching practice can be discussed, from the lesson plan to the resources and from how the trainee delivered instructions, to how the trainee explained new language. A recurring topic of talk, however, is language teaching pedagogy. This requires trainees to understand and even begin to use the jargon of the English language teaching profession, which can sometimes seem confusing, if not opaque.

One of the areas that can be particularly troublesome is appreciating the difference between lessons designed to practise language skills (listening, speaking, reading, writing) and those designed to develop language systems (grammar, vocabulary, phonology, discourse), each of which, according the assessment criteria of this centre's awarding body, requires a different approach. This distinction can be confusing for trainees, as Extract 1 shows. Notice how the trainee's question in lines 28–9 reveals a failure to understand both the association of presentation and practice with a language focus, and the trainer's point about skills development.

Extract 1

TR = Trainer; TE = Trainee

001	TR:	Okay can I just um just a kind of
002		slightly pedantic point? I think
003		it's really nice the way you've
003		just given Simon some feedback,
004		but just in terms of *skills*
005		lessons, you don't have to think
006		about it in terms of presentation
007		and practice....so Simon's last
008		activity was really much more of a
009		kind of follow up activity,
010		personalising what they'd listened
011		to a bit and giving them the
012		chance to kind of get involved...
013		but they weren't practising the
014		language because there wasn't any
015		language that he'd been working on
016		particularly. It wasn't a *language*
017		*focused* session. So I'm only
018		saying this because that happens
019		on plans that's why I'm just
020		picking it up. Just remember when
021		you're having a skills lesson you
022		don't need to think about
023		presentation and practice; you
024		think in terms of the stages of
025		your listening like what you're
026		doing to help them predict the
027		topic.
028	TE:	So what would you call that would
029		you call that? Practice?

030	TR:	You'd call that usually a 'follow
031		up task' to personalise the topic
032		or to give them the chance to
033		respond personally in some way to
034		the topic of the of the listening.
035		So ... you'd normally call it a
036		follow up task really.

Input sessions on training courses will generally cover the vocabulary of language teaching but, for trainees with little understanding of how languages are learnt, the distinction can remain problematic. In the following extract, the trainee has taught what she considers a listening lesson, but has described it as a systems (in this case, grammar) lesson:

Extract 2

001	TR:	Mm, I mean, really you put this
002		lesson as a systems lesson didn't
003		you as a kind of language. The
004		main aim was focusing on language
005		and it felt to me, it felt more
006		like a skills. The principle aim
007		was the skills and that the
008		language bit was an extension you
009		know the the listening took a long
010		part of the lesson
011	TE:	No I wanted to do a listening
012	TR:	Yeah, but I'm talking about on
013		your plan

As with any new language, learners need more than mere exposure to assimilate and use new vocabulary and understand new concepts. Early on in the training programme they need to be clear about how assessments are to be made on their teaching so that they can prioritise the language and concepts they need to learn.

In Extract 1 above the trainer comments on feedback given to a trainee (Simon) by another trainee. This takes us to the second task-at-hand, feedback on the teaching that has taken place. Feedback is generally of two types: positive and negative critique of the lesson, and suggestions and advice about how to improve the lesson. Feedback can be provided by the trainee who has taught (self-evaluation); by other trainees (peer feedback) and by the trainer (trainer feedback).

My interview data revealed that trainers and trainees disagree about the value of each type of feedback. Trainers believe the purpose of the feedback event is to, 'get trainees to evaluate the lessons and develop those critical skills' and 'to help people along in their own development...to give them support in discovering things about their own teaching'. This is a process view of feedback which puts the emphasis on trainee self-evaluation and peer evaluation rather than trainer evaluation.

Trainees, on the other hand, felt that the feedback event should help them 'to develop as teachers' and 'to give guidance on how to improve teaching'. They expected the focus to be on trainer evaluation so that they knew what they had done right and what they had done wrong. They were performance focused and more interested in the teaching product than the teaching process.

This mismatch in 'sharedness' of expectation (Blommaert 2007) of the tasks at hand led to frustration: one trainee summarised his dissatisfaction with the focus on self-evaluation, complaining that:

> For me...activities which are designed to draw people out I personally find a bit tedious and time wasting and I would rather have somebody, particularly the assessor, say, 'this is what I felt about it'.

While another accused a peer of being uncooperative in the peer feedback:

> I felt like you've been really negative a lot of the time and it's made it difficult for me to say anything, because I don't feel like you take it on board if I say anything.

Such a mismatch in expectation, however, is not unique to preservice language teacher education. Acheson and Gall (1997), for example, mirror the view of the trainers here, suggesting that the purpose of the feedback session in a clinical supervision context is to encourage teachers to talk reflectively about their lessons, while Vásquez (2004: 35) argues that those who are observed 'expect to receive from the post-observation meeting...a balance of positive appraisal and constructive criticism' and are less concerned with developing reflective skills.

If trainees are to self-evaluate and provide peer feedback, they need to be aware of this before joining a course. It would be useful, too, to give such trainees clear guidance about how to perform these discourse activities supportively. Providing trainees with examples of successful peer interaction in feedback could also go some way to persuading trainees that they can learn a great deal from each other as well as from the trainer. Trainees can also help by adopting an attitude of openness to new ideas and by accepting that some of the responsibility for exploration must rest with them.

Roles and relationships

Another generic feature of the feedback event is 'roles and relationships'. Trainers have two main roles on initial training courses: to help trainee teachers to develop language teaching skills and to assess teaching against a set of standards, usually provided by the awarding body. It has been suggested that these two roles are incompatible (Holland 2005) as one focuses on nurturing and development, while the other is concerned with assessment and 'gate-keeping', that is, deciding who should join the profession and who should not (Roberts and Sarangi 1999).

Certainly, proficient trainers are aware that they must perform these two roles and are able to do so skilfully. Often it is obvious which role they are realising. In Extract 3, for example, the trainer is asking the trainee to describe the learning outcomes (also often called 'learning objectives' or 'aims') of the lesson, a skill that most training courses will set as a key standard or criterion for trainee teachers:

Extract 3

001	TR:	But we still haven't established
002		what your aims were have we? We
003		still haven't really established
004		what it was that you were focusing
005		on, hoping to help learners with.

The trainer's unambiguous message is that the trainee will need to state and fulfil learning outcomes in order to pass this part of the course.

At other times, it is the trainee's general development that is the focus of the feedback. In Extract 4, the trainer has asked trainees to work on aspects of their teaching which they find particularly problematic. The trainee has identified giving instructions as one such area and so the trainer offers some practical advice about this personal development aim:

Extract 4

001	TR:	I mean if it is something that
002		you're struggling with, and you do
003		feel you want to work on it, one
004		thing you could do is actually
005		write out exactly what you plan to
006		say. It doesn't mean that you then
007		have to learn it off by heart but
008		it's a way of really clarifying
009		the key things that you need to
010		get across.

However, most of the time, trainees endeavour to perform the roles of assessor and developer at the same time, incorporating the assessment criteria within a more developmental discourse. In Extract 5, Clara, the trainer, wants the trainee to think about the value of what she has taught to the group in terms of how meaningful it is. To do this she asks the trainee to put herself in the place of the foreign language learner:

Extract 5

001	TR:	Imagine these are verbs in a
002		foreign language and you made the sentence 'he
003		pretended to be French' 'he
004		attempted to break the record'.
005		When you go out of the room how
006		long would you remember those
007		phrases, do you think? Would they
008		would they really stick in your
009		mind, do you think?

There is assessment in this section: the trainee needs to teach the learners 'meaningful language', that is, language which students can use in their worlds. The trainee has not demonstrated this skill in the lesson she has just taught. The trainer, however, also offers advice that is more wide-ranging, that might help the trainee to improve her teaching generally: to put herself in the place of her learners. While this approach might help her to achieve the assessment criterion, it is not itself measured or assessed: its purpose is developmental.

The trainee must learn to differentiate between comments made that are evaluative and comments that are developmental (see also Hooton, this volume), particularly on short courses where the pressure to meet criteria or standards within a short time-frame is immense. It is not always easy to do so, especially when criteria are not referred to explicitly in the feedback or when the advice given underpins a standard rather than refers to it directly. In Extract 6, the trainer is criticising the excessive amount of 'teacher talk' in the lesson, something which is generally regarded as undesirable (Willis 1996):

Extract 6

001	TR:	The amount of talk that took
002		place; I don't know how you how
003		you felt. It seemed to me that
004		there was a lot of you talking in
005		that lesson, more than I've seen
006		up till now. I don't know whether
007		that was just my impression or did
008		you feel that as well?

It is not uncommon for teacher trainers to focus on this issue in feedback. This is not because 'controlled teacher talk' is an assessment criterion, but because a reduction in teacher talk underpins the common assessment criterion of creating a student-centred class, a class in which students have the opportunity to practise speaking.

This shift to classrooms characterised by student talk can challenge trainee expectations of the role of the teacher. Simon, a trainee, has previously trained as a music teacher, an experience which he believes encouraged a teacher-fronted class. In Extract 7 he discusses the tension he feels exists between his music teacher training and his English language teacher training:

Extract 7

001	TE:	Maybe this is from my background
002		but .. I've got to get into the
003		habit that it's not a cop-out you
004		know .. that is, actually, you
005		know, the focus. That's what we

006	should be doing. We should be
007	giving them that sort of
008	practical. Cos I'm not actually
009	teaching anything, it does feel
010	it's a bit of a cop out

For teachers with other subject specialisms, the shift to language teaching can be as challenging as starting a career in language teaching is for others. That this trainee feels able to articulate his anxiety about the conflict in expectation is testament to the supportive atmosphere his group has created in the feedback event.

Trainees, too, have a number of roles to play in the feedback event. In the role of trainee, they are learning how to teach English and so must listen and take on board advice from others, a role which they expect to play but may not always find easy to play. This trainee, for example, found critique easier to take when it was not so focused on the detail of the lesson:

> I felt more comfortable in the actual feedback sessions since we changed tutors because it's been a more a more general nature, whereas I think it was incredibly detailed in the first ones.

Another trainee found it easier when the feedback event was structured:

> Both trainers have tried more like formulaic or like structured feedback sessions...I think Madeline's division of the board and Lauren's green card, blue card, yellow card systems work really well.

As explained in an earlier section, trainees must also take on the role of the 'trainer' as they feedback to their peers. How successful they are in this role depends on a number of factors, not least the relationships they have formed with others in the group. Where relationships are strong, trainees seem more able to take on the trainer role. In these cases, the atmosphere is highly collaborative and cooperative, and often 'trainees as trainers' are thanked for their feedback. Peer feedback can also be descriptive rather than analytical, and focus on the strengths rather than the weaknesses of teaching practice. When negative peer critique is offered, it is often linked to a weakness the 'trainee as trainer' had with his/her own teaching and can focus on minor aspects of performance, as in this example:

> The only thing for working on, I thought, was the same as what I did last week. When somebody says the wrong answer, I did exactly the same and I said 'Oooh no!' and then in my feedback I think you

suggested I say something like, 'Why did you think that?' or, 'Did everyone else get the same answer?' and so try to draw them out.

In order to avoid either descriptive or trivial trainee feedback, trainers may set observation tasks for trainees. These can focus the untrained eye of the trainee on important features of the language classroom and create a greater degree of criticality in the feedback. Structured tasks can also provide a welcome support to trainees who are in groups where relationships are not congenial, reducing the risk of offending peers with what can be seen as negative critique.

Organisation

Writing of dyadic feedback sessions, Waite (1993) identifies three separate phases: the supervisor report phase, the teacher response phase and the programmatic phase. He demonstrates that those trainees who understand the phases and who work through them in a collaborative way with the trainers, accepting each other's roles, are considered by the trainers to be 'unproblematic' (ibid: 691). On the other hand, trainees who do not allow the phases to unfold, pursuing their agendas in an adversarial way, can be seen by trainers as uncooperative and resistant. Again, a mismatch in understanding of how the feedback session should flow can lead to serious consequences for trainees: in the case Waite reports, the trainer recommends that the trainee is not offered a contract as a teacher.

In my data, different organisational principles hold, depending on the feedback technique the trainer introduces. For example, in one feedback session, the trainees were asked to write positive and negative comments of their own and of their peers' teaching on the whiteboard. In another, trainees were asked to write a question that they wanted to ask each trainee about their lessons. However, most of the feedback sessions included some or all of the phases shown in Table 1.1 and although the order in which they appeared was less than rigid, self-evaluation generally preceded peer feedback, which preceded trainer feedback.

Of course, stages do merge into each other, especially the Questioning Phase. However, in order for the feedback session to proceed smoothly, and for trainees' contributions to be positively evaluated by trainers, trainees must know which phase they are in, understand its purpose and respond accordingly.

When trainees do not contribute to the phases appropriately, for whatever reason, the flow of the feedback breaks down and the trainer can view the trainee in a negative light. In Extract 8, the feedback session is

Table 1.1   Phases in feedback sessions

Phase 1	Trainee is asked to comment on his/her lesson.	Self-evaluation phase
Phase 2	Trainer asks trainee questions about particular sections of the lesson.	Questioning phase
Phase 3	Trainer gives positive and negative feedback.	Trainer feedback phase
Phase 4	Trainer asks the other trainees to comment on the lesson.	Peer feedback phase
Phase 5	Trainer provides a summary of strengths and weaknesses in the lesson.	Summary phase

in Phase 4 – *Peer Feedback Phase* - and the trainee, Stuart, has been asked by the trainer, Ned, to comment on a peer's teaching (Calista):

Extract 8

S = Stuart; N = Ned; C = Calista; J = Josephine

001	N:	Um, (.) Stuart you you've gone a	
002		bit quiet	
003	S:	Sorry about that ((quietly))	
004	N:	Do you want to eh start on	[with
005	C:		[on
006		Calista	
007		((laughs))	
008	N:	On Calista	
009	S:	Umm (…) No ((falling tone)) no	
010	N:	No?	
011	S:	I've I've got some thoughts sorry	
012		but I wasn't making notes terribly	
013		well today and I've got no real	
014		structure to them so I'll have to	
015		generalise from them	
016	N:	If they've got no real structure	
017		to them that's fine	
018	S:	Um	
019	C:	((laughing)) You can write notes	
020		but not necessarily in the right	
021		order	
022		((laughs))	

023	J:	((laughs))
024	C:	(xxxxxxxxxx)
025	S:	The exercise at the end was really
026		interesting fun and could have
027		been longer and

This section begins with a question from the trainer, which seems to criticise Stuart's lack of input. His 'Sorry about that', which is said in a very quiet way, gives no reason for the quietness nor provides the feedback that Ned seems to expect. Nevertheless, Ned pursues Stuart for some feedback when he asks 'Do you want to start on Calista?' Stuart's 'no' in line 8 is not the answer Ned was expecting. In most cases, if asked to provide peer feedback, trainees will do so, even if they have little to say. Stuart's 'no' could be interpreted in at least two ways: it could be that he does not understand the task in hand or his role in it. Alternatively, he could be resisting the role of 'trainee as trainer' through refusing to cooperate. Because he soon changes his mind and begins his feedback (line 25), and because Calista suggests that he has written notes (line 19), the second interpretation is more likely.

Interviews with the trainer, Ned, and the trainee, Stuart, corroborate this view. There was tension between these participants: Stuart was dismissive of some of Ned's feedback and Ned found Stuart 'difficult'. When asked why he said 'No' at this point, Stuart stated that 'I didn't feel I could make any criticisms', a comment which refers to the poor relationships between participants in the group. However, Stuart's stance makes him vulnerable. By not cooperating in the feedback he could be jeopardising the grade he receives and even whether he passes the course.

The 'Questioning Phase', is perhaps the most peripatetic of the phases. It is performed only by trainers and can interrupt self-evaluation and peer-feedback as well as being embedded in trainer feedback and the Summary Phase. The Questioning Phase comprises a set of probing questions which are asked by trainers in order to focus on sections of a lesson which are particularly relevant in terms of what they see as good, but more often bad, practice. If a trainee is unable to respond to the questions, to show understanding of the mistakes made, the result can be uncomfortable and trainees can seem either uncooperative or ill-prepared. In Extract 9, the trainer and trainee are talking about the lack of language work in the lesson:

Extract 9

001	TR:	We haven't really been able to
002		establish what kind of language

003		you were hoping learners would use in
005		conversation. Like I put an
006		example
007	TR:	Yeah, but what kind of natural
008		conversation? Like the natural
009		conversation we're having now or
010		like the natural conversation you
011		have in the break? Or (...)
012	TE:	Well, I said, 'What sort of things
013		do you talk about like a lot?'.
014		One of them said, 'Oh, deciding
015		what to eat on the menu. I think I
016		will have that.'
017	TR:	So you were hoping to develop
018		phrases like, 'I think I'll have
019		this for my starter' or 'What are
020		you going to have?' or is that
021		what you were hoping to develop?
022	TE:	Yeah that they would just sit and
023		chat
024	TR:	And at what stage did you plan to
025		give them help with that
026		particular language in that case?

The trainer is trying to make the trainee understand that she has not given the language part of the lesson enough consideration, but the trainee, at this stage in her development, is either unaware of the expectation that trainees should teach learners new language in a lesson, or is making rash statements in order to provide some kind of answer in response to the trainer's probing. Either way, she is not winning the approval of the trainer.

Such probing by trainers is not, in my experience, designed to embarrass trainees, although doubtless some trainees are made uncomfortable; rather, the process is used to highlight to trainees fundamental features of language teaching that trainees must be able to manipulate in order to pass the course. Like other phases in the organisation of the feedback event, trainees must contribute to the questioning phase and demonstrate that they understand its pedagogic purpose. The most successful trainees do so by engaging with the questioning in a thoughtful and creative way, taking the floor and developing a dialogue with the trainer. As Waite (1993) argues, collaborative relationships are a key feature in successful feedback.

Suitable resources

Language is the key resource in the feedback event. Trainers, in particular, use their language resources to represent their positions and ensure that the feedback event proceeds smoothly and that trainees learn from the experience.

Trainers give positive and negative feedback to the trainees, but giving negative feedback can be difficult. To deliver this successfully, trainers must negotiate the 'face needs' of the trainee. Goffman (1967: 5) defines face as:

> the positive social value a person effectively claims for himself by the line others assume he has taken during a particular contact.

Face, then, is tied up with ego and identity, with threats to face resulting in embarrassment, anger and tears, amongst other things. Brown and Levinson (1987) have explained how face can be threatened in a number of ways, but the Face Threatening Acts (FTAs) most pertinent in the feedback event are suggestions, advice, warnings, criticisms and challenges, all of which trainers might perform in order to encourage trainees to meet assessment criteria and improve their teaching skills. Trainers are aware that all these speech acts could damage the face of trainees and so they use a number of strategies to mitigate their force. Three of the most wide-ranging in my data are eliciting, hedging and claiming common ground.

Eliciting

Eliciting is a questioning technique. It differs from the Questioning Phase discussed above in that it is not used to focus in on particular issues in teaching. Rather, it is a technique used by trainers to organise the feedback and to ensure that trainees stay involved throughout the event. There is another, perhaps less explicit reason, that trainers employ elicitation: it means that problems in the teaching practice are identified by trainees themselves rather than seeming to be a concern of the trainers. This linguistic resource helps trainers to protect the trainee's face. Through identifying teaching and learning issues themselves, trainees can claim 'positive social value' (Goffman 1967: 5) as this ability is positively evaluated in the feedback event. If the trainer has to raise the issue, this opportunity is lost and the discourse act immediately becomes face threatening.

Consequently, much eliciting is general in scope and focus, with phrases such as 'how do you feel it went today?', 'any thoughts?' and 'have you got any comments?' all being common in the data. It is what follows that affords opportunity for trainee concerns to emerge. In my data, discourse that follows eliciting moves such as those listed above is often self-evaluative, detailed and personally relevant to the trainees; in fact in one feedback session, apart from eliciting moves and some back channelling ('mm' 'yeah' and so on) there is very little trainer talk for the first five minutes of the feedback event.

Trainees can prepare to respond to elicitation moves by trainers through taking notes during and after the teaching practice and by listing positive and negative features of their own and their peers' teaching. The trainees in my data who were involved in this way responded positively to elicitation and were viewed as cooperative by their trainers.

Hedging and claiming common ground

Hedging is the use of language to soften the force of an utterance. There are a number of ways that this can be done, for example, through modal verbs (*could, might*, etc.), lexical modals (*perhaps, maybe, possibly, suppose, think*, etc.), and through intensifying adverbs (*sort of, kind of, quite, rather*). Trainers use hedges frequently. In this example, for instance, the trainer is responding to a concern of the trainees that the group they are teaching is of a much lower language level than the previous group. The trainees have suggested slowing down their delivery as a way of coping with the level, but the trainer does not agree with this approach:

> *Possibly thinking* about that, but also it *might be* worth *thinking* about how else you can give them extra support, so *maybe* now that we know who it is that's an elementary learner, *maybe thinking* about where they're placed in the classroom or once you get to know the learners *a bit* better you can *think* about who, who *might* be able to help them with different things.

Each of the suggestions the trainer makes to counter that given by the trainee – giving extra support, seating arrangements and peer support – attracts modality: lexical modals (*possibly, thinking, maybe*); verbal modals (*might*); and an intensifying adverb (*a bit*). So while this utterance is full of potential FTAs – one disagreement and three suggestions – their force is reduced by the amount of mitigation the trainer packs around them.

This pattern of modal use to reduce the FTA potential is a common feedback technique of trainers. In the following excerpt, the trainer wants the trainee to put students into pairs rather than into groups of three when they check their answers to a listening activity:

> I *think* it's, I *probably*, I *mean* threes is okay but *probably* pairs is your ideal for comparing answers to a listening or something cos it's least threatening you know when you you *can really* feel stupid sometimes *can't* you I dunno in a class well in any class *probably* but in a language class because *sometimes* you can't understand something you *tend to* feel *a bit*, it *can* make you feel *a bit* stupid so having to *kind of* reveal that to more people is *a bit* more threatening so I *think just* sharing it with a partner is *just probably* the least threatening way of doing it.

The trainer is careful to ensure that her one FTA – offering advice – is as unthreatening as possible through linguistic hedging and through another device to reduce the effect of FTAs, that of claiming common ground. Claiming common ground means that the speaker aligns him or herself with the 'goals and values' (Brown and Levinson, 1987: 103) of the hearer: in this case, the trainer claims this ground through the use of the inclusive 'you' and through suggesting that all language learners (herself included) 'can feel stupid' if they cannot understand something. Nevertheless, the amount of mitigation means that main message – put the students in pairs to check their answers, not threes – takes over five times longer to deliver.

For a trainee, mitigation can make negative feedback bearable:

> You want the negative delivered in a gentle, smiley, nice kind of 'we're with ya' kind of way...It's not so much the criticism, it's the way it's said in terms of treading on each other's toes, or whatever. Like there's a way to give criticism that sounds positive and constructive and there's a way that makes it sound just like criticism.

However, other trainees can feel that hedging is not always necessary:

> I think a lot of time is wasted just by people not wanting to tread on so many toes and things have been like maybe too sugar-coated, you know?

Furthermore, the cumulative effect of the hedges coupled with other approaches, such as the claiming common ground technique, can make the key message difficult to access. Vásquez (2004) suggests that 'supervisors' (trainers in this context) can be so concerned with protecting the face needs of the teachers they have observed that their advice is lost or buried in other talk. This leaves the teachers feeling short-changed or unable to recall what advice they did receive.

Conclusion

The feedback event in initial English language teacher training is a complex genre which demands a great deal from those who participate in it. Trainers usually have the luxury of observing feedback before having to perform it and then have multiple opportunities to develop their role in this communicative event. Trainees, however, do not usually have these opportunities. They must learn their roles as they perform them and follow the tasks on hand at the same time as they take part in them. Failure to understand or to participate in the feedback according to the 'conventionalised expectations' can lead to negative evaluation by the trainers, which can eventually affect the grade the trainees are awarded.

To avoid this situation arising, prospective trainees should be thoroughly briefed on the generic features of the feedback event before joining the course. Sections of data, such as those presented here, can be used to illustrate these features and to demonstrate the ways in which trainees can appear as cooperative or uncooperative. Finally, the philosophy underpinning the approach to feedback should be explained so that trainees can decide independently whether they wish to engage in the feedback genre and to what extent. These precautionary activities should enable both trainers and trainees to engage profitably in feedback from the start and should limit the chances of 'sharedness of expectation' being unsuccessful.

Trainees, then, must be aware that beneath the hedges, softness and claiming common ground, trainers are making serious points about teaching and learning that trainees will need to address in subsequent sessions. Recognising this message, however, may not always be easy as trainees may be confused about either the content of the message or its importance. Although trainers will often summarise the main points of the feedback and will, in most cases, provide written feedback too, trainees should ask for clarification if they are in the least unclear. Trainers, too, need to be aware that mitigation can cloud the message or reduce its force, a reality that needs to be considered when feedback is linked, particularly to assessment.

Trainees can also be encouraged to make notes in the feedback sessions so that important feedback is not lost in the mass of talk that feedback generates. This is just as important when feedback is directed at other trainees in the group. As one of the trainers interviewed said:

> Often there are similar issue that are coming up in people's lessons, and sometimes you know, it's that thing of learning peripheral learning, isn't it, I suppose? You know, when the focus isn't on your bit when you are able to think about things more deeply. Perhaps when you're actually thinking about your own lesson you think about it less, in a way, because there's a sort of pressure.

References

Aechson, K. A. and Gall, M. D. 1997. *Techniques in the Clinical Supervision of Teachers: Preservice and Inservice Applications* (4th edn). New York: Wiley & Sons.

Blommaert, J. 2007. 'Genre and the asylum interview'. Talk given at Ethnography, Language and Communication, London, Institute of Education.

Brown, P. and Levinson, S. 1987. *Politeness: Some Universals in Language Use.* Cambridge: Cambridge University Press.

Creese, A. 2008. 'Linguistic ethnography'. In A. Creese, P. Martin and N. Hornberger (eds), *Encyclopedia of Language and Education*, Volume 9. New York: Springer.

Holland, P. 2005. 'The case for expanding standards for teacher evaluation to include an instructional supervision perspective'. *Journal of Personnel Evaluation in Education*, 18(1): 67–77.

Goffman, E. 1967. *Interaction Ritual: Essays on Face-to-Face Behaviour.* New York: Doubleday Anchor.

Rampton, B. 2006. *Language in Late Modernity: Interaction in an Urban School.* Cambridge: Cambridge University Press.

Rampton, B. 2007. 'Neo-Hymesian linguistic ethnography in the UK'. Special Issue *Journal of Sociolinguistics*, 11(5): 584–607.

Roberts, C. and Sarangi, S. 1999. 'Hybridity in gatekeeping discourse: Issues of practical relevance for the researcher'. In C. Roberts and S. Sarangi (eds), *Talk, Work and Institutional Order Discourse in Medical, Mediation and Management Settings.* Berlin: Mouton de Gruyter.

Vásquez, C. 2004. '"Very carefully managed": Advice and suggestions in post-observation meetings'. *Linguistics and Education*, 15: 33–58.

Waite, D. 1993. 'Teachers in conference: A qualitative study of teacher-supervisor face-to-face interactions'. *American Educational Research Journal*, 20(4): 675–702.

Willis, J. 1996. *A Framework for Task-based Learning.* London: Longman.

2

The Design of Post-Observation Feedback and Its Impact on Student Teachers

Nur Kurtoglu Hooton

An extract

N = Nur; J = Jake; L = Lisa

001	N:	... So a very competent use of
002		materials overall. Overall good.
003		((checking notes))
004		Oh yes, when you finished the
005		lesson, do you remember what you
006		said, at the end of the lesson?
007		(.) You said "for me – that's my
008		lesson over."
009		((L breaks into laughter))
010		... It was like you were so
011		relieved. ((addressing whole
012		group)) What are some ways of
013		finishing off a lesson?
014	J:	You would like to get <u>rid</u> of <u>me</u>
015	N:	I'm all done now
016		((laughter from whole group))
017	N:	What would be one way of finishing
018		off?
019	J:	Quite good I think just to sum up,

020		you know, what you have done just
021		very quickly.
022	N:	So wrap it up by either saying to
023		them what you've been looking at
024		or get them to say what they
025		believe they've done.

Exchanges as in the extract above would be immediately recognisable to any student teacher who has received feedback on her/his teaching practice. Teacher educators expect these teachers to do teaching practice and they themselves observe the teaching, and give feedback – to the whole group and/or to individuals. They also expect student teachers to reflect on the feedback that they have received. This is all part of a recognisable framework for teaching practice observation. But what about the possible *effect* of the feedback on those teachers?

In this chapter, we shall first explore concepts and issues surrounding post-observation feedback. We shall then look at feedback encounters and we shall see what several student teachers themselves say about the effect of such encounters. As Freeman states:

> While we might arrive at crudely accurate maps of teaching by studying it from outside in, we will not grasp what is truly happening until the people who are doing it articulate what they understand about it.
>
> (2002: 11)

We shall also explore the role of post-observation feedback as an instigator of teacher learning and change.

A journey into learning to teach

Feedback on teaching practice forms an essential part of many teacher training courses, including those at initial teacher training levels. It is through observed teaching practice that beginning teachers are provided with opportunities 'to get a feel' for classrooms and for becoming a language teacher. It is through tutor and peer feedback that they are encouraged to develop and progress.

The programme, which forms the focus of this chapter, is delivered at a British University. It is a 100-hour, four-week, initial teacher training programme intended for those who have little or no teaching experience.

It is designed to help the beginning teachers 'develop an understanding of theoretical ideas created by others while simultaneously beginning to develop their own personal theories of learning and development' (Kroll 2004: 201).

Candidates come onto the programme as individuals with their own prior learning experiences, their views on teaching and learning, their own beliefs and aspirations.

During the programme, the teachers are provided with the opportunity to observe English language classes at a local language institution. These classes serve as initial input for the student teachers. Seeing experienced teachers in action and reflecting on this experience through some observation tasks, the teachers are encouraged to investigate principles that underlie effective teaching and to transfer some of the practices into their own teaching.

Following their initial teaching experiences, post-observation feedback sessions provide an ideal opportunity for the student teachers and the teacher educator/s to jointly create knowledge and meanings in the pursuit of learning and improving teaching.

Group feedback sessions

Post-observation feedback on the programme is held in groups and where the tutor considers it appropriate, individual feedback also follows. The usefulness of group feedback sessions cannot be denied. The following extracts from student teachers speak for themselves:

> It is useful to have the opportunity to voice our concerns about teaching to an experienced teacher as well as to others in our position. Hearing different opinions helps us to widen our knowledge and outlook.
>
> (Marie, August 2003; author's data)

> I am just glad that we have these periods of feedback because without the three perspectives it is difficult to know what to improve on; ie self-evaluation, tutor evaluation, peer feedback.
>
> (Sam, August 2003; author's data)

Post-observation feedback, which requires many skills and techniques, is a well-researched area (see, e.g., Edge 1993/1994; Freeman 1982 and 1990; Gebhard 1984; Kumaravadivelu 1999; Kurtoglu-Eken 1999; Randall and Thornton 2001; Roberts 1998; Wallace 1991 and 1999).

Research has also been done in the areas of how teachers develop professionally and how they change their practices. Gebhard (1990: 118) sheds light on how 'opportunities are provided (or blocked) for beginning teachers to change their teaching behaviours'. His study provides evidence of change teachers experience through 'multiple activities'. What the present chapter does is to focus more specifically on the possible effect post-observation feedback, in particular, may have on the way student teachers learn and change.

Kinds of feedback

In order to understand feedback and any possible learning or change teachers may experience, we can usefully draw on research conducted not only into observation feedback but also in many other fields. Korthagen and Russell, for example, maintain that teacher educators 'require knowledge, skills, and attitudes in the field of human development (adult development, social psychology, counselling, and the like)' (1995: 191).

Particularly relevant within the context of giving feedback in teacher education is Egan's work (1990, 2002) on counselling. Egan maintains that 'feedback is one way of providing both support and challenge' (2002: 360–1) and argues that clients would need to know how well they are performing if they are to be successful in implementing any action plans. He identifies two kinds of feedback which he calls *confirmatory* and *corrective*.

Adapting the above terms for my purposes, I define them as follows: Confirmatory feedback involves positive feedback in the form of praise, or confirmation and/or reassurance that something went well. This 'something' can involve a teaching skill, a teacher quality, some teacher behaviour, or even a decision the teacher may have taken during teaching practice. Corrective feedback, on the other hand, applies to situations where there was perhaps a better alternative for some skill that had been exhibited, for some behaviour that took place, for some teacher quality that was or was not revealed, or for some decision that did not work particularly well. In essence it acts as a 'correction' while confirmatory feedback provides 'a pat on the back'.

For assessment purposes, to ensure that student teachers have fulfilled the aims and objectives of an initial teacher training course, certain behaviours are expected, and therefore feedback is often likely to involve feedback of a corrective nature. Feedback which specifies, for example, the need to show awareness of the learners' errors and be able to correct them sensitively, or the need to ensure that there is a purpose for using an activity, involves corrective feedback. Student teachers might feel that this type of feedback might also involve some kind of 'a gentle telling off', especially if the feedback implies that there is very little or no evidence of progress.

Feedback is not always corrective, however. There is also the kind of teacher educator feedback which focuses on what is good, what can be celebrated, what it is that worked well: what can be called confirmatory feedback.

Bullough Jr. and colleagues discuss how teachers, and especially student teachers, are 'vulnerable to criticism' and 'even the best educated and most able emotionally secure of beginning teachers face moments of frustration and self-doubt' (1991: 79). It is the teacher educator's responsibility to ensure that feedback is discussed in a safe and appreciative environment.

Realms of feedback

It was noted above that Egan's work on counselling is relevant within the context of post-observation feedback. Work done in organisational behaviour by Senge (2002) is also relevant to our discussion here. Senge explores the way learning organisations behave and describes the meaning of profound change in the context of such organisations:

> ...we use the term 'profound change' to describe organizational change that combines inner shifts in people's values, aspirations, and behaviors with 'outer' shifts in processes, strategies, practices, and systems....In profound change there is learning.
>
> (Senge 2002: 15)

The references to inner and outer shifts in these words from Senge are particularly significant concepts as they have strong implications for post-observation feedback sessions. In organisations, companies and institutions, change can be introduced intentionally by managers – this change can be implemented top-down, or with collaborative efforts (with the help of the affected group). In educational contexts such as the post-observation feedback sessions that we are concerned with, the teacher educator can be seen to function more as a facilitator and an instigator of some of the learning and change that may take place during the training course and beyond.

In a feedback-giving context, the teacher educator functions in the realm of the outer shifts. A teacher's peers giving feedback during a post-observation feedback session also function at this level. It is the responsibility of the teacher receiving the feedback, however, to combine the outer realm with their own values, aspirations and behaviours (inner realm) in order to move forward.

In the present context, the outer realm would encompass all the 'observable' aspects of a teaching practice session, as well as the 'application' side of things: how a teacher exhibits their teacher persona, teaching methods s/he uses, and classroom skills, timing and pace, use of materials and, last but not least, learner reactions; in sum, all the criteria that go into a lesson observation form that a teacher educator may use to assess a lesson. The 'inner realm' in the context of teaching and teaching practice would involve more abstract concepts such as the teacher's learning experiences, her/his aspirations in teaching, as well as beliefs in teaching and learning. Any reflection that may take place before, during and after teaching brings the outer and the inner realms together.

Figure 2.1 explains how it might be possible to conceptualise and illustrate the relationship between the inner and the outer realms.

The teacher educator's feedback, which operates at the outer realm (but which may also challenge the inner realm), is fed into the teacher's inner realm. Here a 'naming' and 'framing' (Schön 1983) process takes place. Feedback received and/or discussed during a post-observation feedback session is matched against the inner realm concepts. This process may sow the seeds for profound change which becomes visible in the outer realm. However, the process does not stop there as the intricate relationship between the inner and the outer realms is likely to be a cyclical process in itself. That is, as the teacher develops s/he is able to feed the feedback into the inner realm and match it against own experiences and beliefs, then put into practice what it is that has been worked on. For there to be changes at the outer realm, there need to be changes in the inner realm. This could apply to corrective as well as confirmatory feedback.

If a teacher educator gives corrective feedback on a teacher's classroom behaviour or classroom procedures, the teacher concerned would be expected to modify their teaching in line with the feedback received. In a way, the tutor would be imposing the change – a change the student teacher needs to make in their effort to learn teaching. In order to exhibit the development needed in their teaching, the student teacher

The Inner realm   The Outer realm

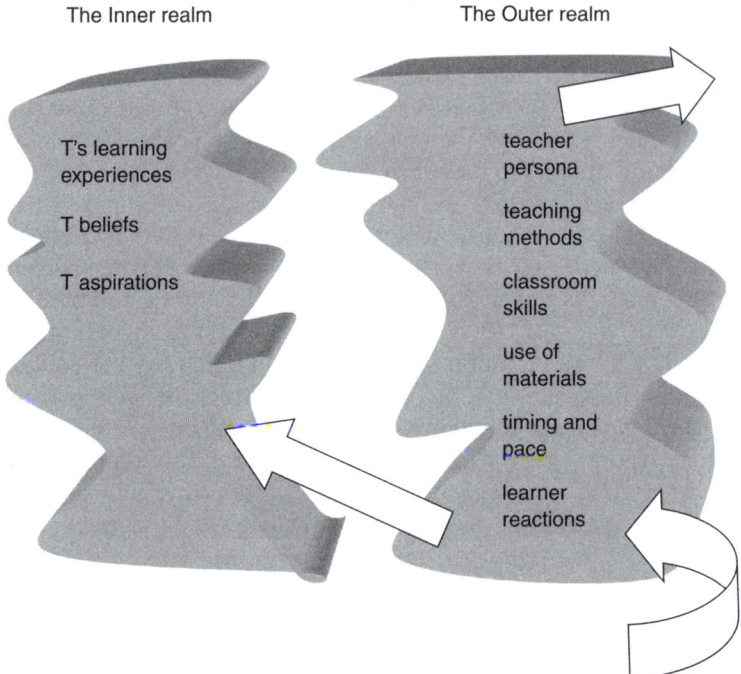

T's learning
experiences

T beliefs

T aspirations

teacher
persona

teaching
methods

classroom
skills

use of
materials

timing and
pace

learner
reactions

Figure 2.1   Inner and outer realms

will have to take the responsibility for reframing the feedback in the inner self.

In instances where confirmatory feedback is received, there is no imposition of change from the tutor. The confirmatory feedback is likely to provide an opportunity for the teacher to explore her/his teacher persona, and to dig deeper down into the inner self. This will help her/him work at a level beyond that of classroom routines and procedures.

Reflection

Reflection is a key term in the dynamic nature of the outer and the inner realms of feedback. Schön (1983) distinguishes between 'reflection-in-action' – the thinking on one's feet – and 'reflection-on-action' – the systematic thinking back on action that has been completed. This distinction is useful within the context of the diagram proposed above in the sense that teacher educators expect student teachers to exhibit the characteristics of a reflective teacher. It is beyond the scope of this chapter to explore reflection and its characteristics in depth. However, assuming that the right conditions for reflection have been established on an initial teacher education course, the reflective student teacher would be expected to draw not only on reflection-in-action but on reflection-on-action, too. In using reflection-on-action the teacher is likely to dig into the inner realm to be able to move to a new level of self-awareness.

Daudelin's definition of reflection, too, serves as a useful guideline for us:

> Reflection is a highly cognitive process. When a person engages in reflection, he or she takes an experience, and filters it through personal biases. If this process results in learning, the individual then develops inferences to approach the external world in a way that is different from the approach that would have been used, had reflection not taken place.
>
> (1996: 70)

Learning may be exhibited in the way a new and different approach to the external world, or in our case, to the teaching itself has been exhibited. And as Wideen and colleagues state 'changing one's teaching is a learning process which involves, in part, building upon and changing prior beliefs and actions about teaching' (1996: 187).

Corrective feedback and teacher learning and change

Corrective feedback requires a period of time for the teacher to process, digest, reflect upon and come to terms with the 'criticism' involved. Change as a result of this kind of feedback is likely to be *convergent* in essence; that is to say, the teacher who receives the piece of corrective feedback will be required to move closer to some agreed norm or form of practice. The teacher is encouraged to exhibit what s/he has learnt. In other words, some observable behaviour is needed for there to be some indication of teacher learning – certainly for assessment purposes.

In a feedback session, realising that the issue of giving clear instructions was not raised by the teachers themselves, I provided the group with corrective feedback and made suggestions as to how they can improve this aspect of their teaching. This corrective feedback seemed to have been significant for three of the four teachers on completing the homework titled 'Learning from TP feedback'. This piece of homework involves the teachers listening to the audio recording of the full feedback session and choosing a one- or two-minute part of the session which they find significant in some way. After transcribing the extract they are asked to comment on why they found the extract significant. On this occasion, three of the four teachers in the teaching practice group had chosen an extract on the part where I focus on 'giving instructions'. Here is what Lisa had to say after she provided a short transcribed extract:

> I chose this extract as it focused on giving instructions to students. It was evident that all of us had difficulties in this area. Consequently, the students had difficulties in repeating the instructions...I realised that if students are going to understand instructions I have to simplify my language. This will help to reduce confusion...clear instructions will also save time during the lessons, allowing time for other activities which can enhance the students' learning....I have been teaching English to students and have never put any thought into how I am going to give instructions....With foreign students you have to think about the content as well as how you are going to word instructions to make it easier for them to understand. Therefore, by writing down the instructions initially and simplifying them it will make it easier for you and more importantly for the students.
>
> (Lisa, August 2003 data)

These comments are in line with the concerns novice teachers seem to exhibit. Livingston and Borko (1989), for example, point out that in their study on expert–novice differences, novice teachers' post-lesson reflections were more concerned with these teachers' 'own teaching effectiveness' rather than learners' understanding of the material which they found to be attributed to 'expert' teachers. The extract from my data shows Lisa's keenness to develop her instruction giving skills, and grading her teacher language. These areas fall under teacher's behaviour, and as Cheng and colleagues (2001: 199) emphasise, the cognitive change involved to advance from novice to expert status is a pedagogical knowledge development which moves from a focus on teacher's behaviour to learner behaviour.

Lisa's comments may exhibit novice teacher behaviour. Interestingly, however, the comments also show evidence of reflection (in this case, reflection-on-action) which is not usually attributed to student teachers.

Korthagen and Wubbels (2001: 133–7) present a useful list of character-istics of reflective teachers. Of the four attributes they list, 'Attribute 4', reads:

> A Reflective Teacher Can Adequately Describe and Analyze His or Her Own Functioning in the Interpersonal Relationships with Others.
>
> (Korthagen and Wubbels 2001: 137)

In the data extract above, the student teacher, Lisa, is using reflection-on-action and clearly shows she is capable of elements of Attribute 4 mentioned here. The extract shows Lisa learning from a corrective piece of feedback she and her group had received. However, learning and change do not have to be initiated by a problem. Initiation could well involve the appreciation of what a student teacher does well. We shall now move on to discuss confirmatory feedback and its potential effect on student teachers.

Confirmatory feedback and teacher learning and change

Confirmatory feedback during a TP feedback session is likely to be given in the context of praise – in connection with what it is that the begin-ning teacher did well. In this context, the confirmatory comments from the teacher educator or peer/s are likely to encourage each teacher to construct his or her own understanding of what teaching involves. As a result of confirmatory feedback the teacher may remain as he/she is; or may also be encouraged to try new avenues and to pursue new chal-lenges. If the latter route is taken, elements of a change that is *divergent* in essence are likely to emerge. Examples will be discussed below.

At the end of each training course, as part of the end-of-course ques-tionnaire the teachers are invited to think back to all the six TP (teach-ing practice) feedback sessions they have had. They are asked to choose one piece of feedback they received (from tutor or peer) that they found significant in some way. They are then asked to comment on why they found it particularly significant and what effect the piece of feedback may have had on them. While the majority of the candidates choose a piece of corrective feedback as having influenced them in some way, it is rewarding to see that quite a few select a piece of confirmatory feed-back to explain how this has been significant for them. Here I will share three vignettes. In the first one, we find that the teacher has been praised on a particular teaching skill – that of 'timing'. He comments on this praise and focuses on its implications for his future involvement of teaching a presentation skills course:

> I was pleased to be praised on my timekeeping abilities in most TP feedback sessions.
>
> I think this showed me that I'm a good estimator of how long an activity will take. This skill is transferable to academic lecturing and presentations too. Thus I will be able to apply it in a variety of class-rooms/lecture theatres.
>
> (Gary, August 2005 data)

In the next vignette, the teacher reports on having been praised for his teacher qualities:

> It was TP 3 (sorry can't quite remember what I'd taught!) but having felt that there were various aspects of the lesson when I might've handled things better, I was really surprised to get such positive comments particularly about what 'type' of teacher I had come across as.
>
> This encouragement made me feel much more confident that things were developing OK and that I was demonstrating attributes of the kind of teacher that I hope to be.
>
> (Sam, August 2003 data)

Here the teacher remembers the praise he received on the kind of teacher he is, and finds this more memorable than the lesson he had taught.

The third vignette is explained in a little more depth, the reasons for which will soon become clear. Some contextual piece of information is first needed.

The extract involves Jake – a candidate who was 53 years old when he joined the 2003 group. One of 12 members in the group, he was the oldest as the average age was approximately 30. Like every course participant, Jake taught six sessions. His initial teaching practice group consisted of Sam, Lisa, Marie and himself.

The extract chosen here is from the feedback on his second teaching practice lesson. The activity under discussion is one which involved him grouping the learners. He had given each learner the telephone number of a fire station, police station, ambulance, or hospital, and invited them to mill around asking one another the question 'Can I have your phone number please?' All learners who had the number for a hospital were then grouped together. So were those who all had a number for a police station, and so on. This grouping worked well, both as a technique in itself and also in terms of its contribution to the overall theme of the lesson which was telephone interactions.

In the extract below, Jake receives confirmatory feedback not only from me as his tutor, but also from all his peers. References such as *You're so natural* (line 1); and *You should write a book about classroom management techniques* (lines 25–26) refer to Jake himself – his teacher persona and what he is perceived as being able to do. Jake is also praised on the success of his activity. Comments such as *It was just excellent* (line 15); *it's a lovely idea* (line 16); *the idea's wonderful* (lines 22–23) all refer to Jake's technique of grouping the learners.

Extract 2

M = Marie; L = Lisa; J = Jake; N = Nur; S = Sam

001	M:	You're <u>so</u> natural.
002	L:	Yeah (4.0) I was saying to Marie
003		earlier, that to think about
004		cutting up the, I don't know
005		whether you thought about that
006		yourself or whether it came from a
007		book I don't know.
008	J:	No, no, it wasn't. I=
009	L:	=It wasn't, no, cutting up the
010		telephone numbers, getting them to
011		meet with each other, and then to
012		do a different=
013	N:	[Hmm]
014	S:	[Hmm]
015	L:	I thought it was just <u>ex</u>cellent!
016	N:	Yes, it's a <u>love</u>ly idea. The only
017		thing was that it actually
018		took longer than it should have.
019	J:	<u>Yes.</u> ((instant agreement))
020	N:	Because of the instructions – it
021		comes down to the instructions as
022		you were saying, whereas the
023		idea's <u>won</u>derful.Very creative.
024	M:	Yes. You're going to get=

025	N:	=You should write a book about
026		classroom management techniques,
027		((laughter from whole group))
028		grouping activities. Really!
029		It's lovely, yes.
030	M:	Its (variance) as well.
031	N:	( ) That was Claire's
032		activity, but he'd adapted it.
033	L:	<u>exactly.</u>
034	M:	It's wonderful.

This extract may look insignificant at first sight as it involves the type of confirmatory feedback that many student teachers may receive in any group feedback session. However, we can appreciate its significance when we explore the impact the feedback has had on Jake. This is what Jake wrote in response to the feedback-related question in the end-of-course questionnaire:

> This may seem minor BUT to be told I am 'creative' has had enormous effects! This creativeness hasn't obviously just happened. However in the past probably because of others' feedback, I would have described myself as a 'bit of a plodder' who needed permission to do anything out of the mainstream. To find I am perceived creative by people has been a bit of a 'life changer'.

His choice of vocabulary – phrases such as *may seem minor BUT* and *has had enormous effects* – is a clear indication that all the feedback Jake received about himself as a person, and about the success of his grouping technique in that particular TP feedback session, seem to have had a transformative effect on him. After all, he confirms this with the final sentence he uses: *To find I am perceived creative by people has been a bit of a 'life changer'*.

Regular correspondence with all the teachers in the group encouraged me to explore Jake's comments further. When I emailed him to ask about the ways in which the feedback may have been a 'life changer' for him, this is what he wrote:

> I have always been perceived as a 'Plodder'. 'Want something doing?' ask Jake, he'll stick at it, get it done and it will be done in the 'correct' way, not quick, because he sticks to 'the rules' is probably how

I have always been seen. Well that is my perception of how I have been seen!

To be suddenly, and it was sudden to me! described as 'Creative' not once but a number of times, and not by just one person (honest!) changed the whole way I think about myself. It has not only restored some of my self-respect but also given me confidence to actually put forward ideas, and opinions, that in the past I would have kept to myself. I would have kept them to myself on the basis that I was a 'plodder' and plodders' ideas whilst not worthless are never new, inventive, creative etc. etc. so not worth airing, who would listen anyway! The 'Creative' comment has also given me confidence to try out new ideas, whether I am confident they will work or not. Something I would have been loath to do prior to the course.

In his response Jake provides some information on how he believes he was always seen by others: a 'Plodder' – a word he visits three times in his response. He does add the comment that that is *his* 'perception of how [he has] been seen!'. While this may indicate that he feels his perception may have been wrong, his comments that follow highlight the significance of the feedback he received from his peers and tutor during the second teaching practice session. He uses the capital letter C for the word 'creative' and puts it into inverted commas, too – both indicators of how significant it is for him to be seen as *'Creative'*. Jake seems to refer to *Someone who is Creative* as someone who is the total opposite of *a Plodder*. He states that the piece of feedback he received on his creativity has 'changed the whole way [he thinks] about himself', that it has *'restored some of [his] self respect'* and given him *'confidence to actually forward ideas, and opinions, that in the past [he] would have kept to [himself]'* on the basis that he had been seen as a 'plodder'. He also adds that it has given him *'confidence to try out new ideas'*, whether he is confident they will work or not.

These comments are evidence that confirmatory feedback on some of his behaviour in class – ones that were seen as creative – provided opportunities for him to gain in confidence, bringing about changes in him as a person. These changes are likely to have been divergent in nature. As his self-image improved, so did his willingness to try alternative ways of doing something. Change may not be immediately observable. When it is, it may manifest itself at different levels, as change can be in behaviour, in conceptions or in the person himself. Jake no doubt experienced changes in behaviour and in conceptions as a result of the confirmatory and the corrective feedback he was given during the post-observation feedback sessions. It is, however, the changes he experienced as a person as a result of the confirmatory feedback that present a powerful image to teachers and teacher educators.

The editors rightly state in their Introduction that:

> It is necessary to look at discourses of post-course experience for evidence of more deep-rooted change that goes beyond conforming to the expectations of the course.
>
> [page xxii]

Indeed, while the above data extract is powerful in its own right one needs to get further evidence to ascertain that the change is more deep-rooted than it may initially seem.

The following extract comes from an email I received from Jake seven months after the programme ended. I had emailed him a copy of a teacher newsletter article I had written (Kurtoglu-Hooton 2004) and as he featured in it, albeit with a pseudonym, I wanted to check that my interpretations had been correct. He wrote a long response volunteering some further information about his 'new found confidence' as he calls it:

> Spooky you should write today. Have just been practicing my classroom management skills on a group of hard-nosed Border Army Guards. We have been given the task of turning them from soldiers to Police Officers inside three months! With a wave of the hand the... Government decreed that the borders would be manned by Police and not Army....I used pictures of vegetables to divide them up....It was great to see the change in attitude once I had attributed the various vegetables [*sic*] characteristics to each group (Peppers group, hot stuff, Cool Cucumbers, etc.) I am not showing off, just another example of new found confidence! 'You did what? Made these hot headed...blokes with guns get into groups by choosing pictures of vegetables!' Believe you me they loved it!
>
> (Jake, March 2004)

As was discussed above, with reference to Figure 2.1, individual reasoning takes place at the inner realm and feeds into the outer realm. If the starting point is confirmatory feedback, this would then act as a trigger for positive emotions during the framing process. If the initial trigger/s were confirmatory, the change that may become visible in the outer realm is likely to be change that is divergent in its essence. This is what Jake seems to have experienced through the feedback he was given by his tutor and peers. He uses his classroom management skills in a way that may seem daring or even inappropriate in some teaching contexts. Yet the fact that he ventures into the grouping technique he used with the police officers shows his confidence and his willingness to test and see what happens. These are no doubt characteristics of a divergent rather than convergent change and are the result of confirmatory rather than corrective feedback.

Final comments

In discussing confirmatory feedback I have intended to highlight the significance of a kind of feedback that is often neglected in the post-observation feedback discursive encounters.

Corrective feedback no doubt provides a powerful learning medium for teachers. Egan (2002: 303) warns us, however, that corrective feedback is all too often very detailed while confirmatory feedback is 'perfunctory'. If confirmatory feedback is detailed in the same way corrective feedback often tends to be, and if it is supported with specific examples from the teacher's lesson, there is every reason that this type of feedback, too, would facilitate teacher learning and change.

I shall leave you with a quote from Ghaye:

> Failures are only one kind of motivation for change. Another is to learn from successful and current 'best' practice.
>
> (2005: 178)

References

Bullough Jr., R. V., Knowles, J. G. and Crow, N. A. 1991. *Emerging as a Teacher.* London and New York: Routledge.

Cheng, Y. C., Mok, M. M. C. and Tsui, K. T. 2001. *Teaching Effectiveness and Teacher Development: Towards a Knowledge Base.* Amsterdam: Kluwer Academic Publishers.

Daudelin, M. D. 2001 [1996]. 'Learning from experience through reflection'. In J. S. Osland, D. A. Kolb and I. M. Rubin (eds), *The Organizational Behavior Reader.* London: Prentice Hall.

Edge, J. 'A framework for feedback on observation'. *IATEFL TT SIG Newsletter,* No. 10 (Winter 1993/1994): 3–4.

Egan, G. 1990. *The Skilled Helper: A Systematic Approach to Effective Helping*. Pacific Grove, CA: Brooks/Cole Publishing Company.

Egan, G. 2002. *The Skilled Helper: A Problem-Management and Opportunity-Development Approach to Helping*. Pacific Grove, CA: Brooks/Cole Publishing Company.

Freeman, D. 1982. 'Observing teachers: Three approaches to in-service training and development'. *TESOL Quarterly*, 16(1): 21–8.

Freeman, D. 1990. 'Intervening in practice teaching'. In J. C. Richards and D. Nunan (eds), *Second Language Teacher Education*. Cambridge: Cambridge University Press.

Freeman, D. 2002. 'The hidden side of the work: Teacher knowledge and learning to teach: A perspective from North American educational research on teacher education in English language teaching'. Review Article. In *Language Teaching*, 35: 1–13.

Gebhard, J. G. 1984. 'Models of supervision: Choices'. *TESOL Quarterly*, 18(3): 501–14.

Gebhard, J. G. 1990. 'Interaction in a teaching practicum'. In J. C. Richards and D. Nunan (eds), *Second Language Teacher Education*. Cambridge: Cambridge University Press.

Ghaye, T. 2005. 'Reflection as a catalyst for change'. *Reflective Practice*, 6(2): 177–87.

Korthagen, F. and Russell, T. 1995. 'Epilogue – teachers who teach teachers: Some final considerations'. In T. Russell and F. Korthagen (eds), *Teachers Who Teach Teachers: Reflections on Teacher Education*. London: Falmer Press, pp. 187–92.

Korthagen, F. and Wubbels, T. 2001. 'Characteristics of reflective teachers'. In Korthagen, F. (ed.), *Linking Practice and Theory: The Pedagogy of Realistic Teacher Education*. Hillsdale, NJ: Lawrence Erlbaum Associates.

Kroll, L. R. 2004. 'Constructing constructivism: How student-teachers construct ideas of development, knowledge, learning, and teaching'. *Teachers and Teaching: Theory and Practice*, 10(2): 199–221.

Kumaravadivelu, B. 1999. 'Theorising practice, practising theory: The role of critical classroom observation'. In H. Trappes-Lomax and I. McGrath (eds), *Theory in Language Teacher Education*. London: Longman.

Kurtoglu-Eken, D. 1999. 'The power of trainer language in training and development'. *IATEFL TT SIG Newsletter*, No. 23: 32–8.

Kurtoglu-Hooton, N. 2004. 'Post-observation feedback as an instigator of teacher learning and change'. Featured article. *IATEFL Teacher Trainers and Educators SIG e-Newsletter*, No. 2.

Livingston, C. and Borko, H. 1989. 'Expert-novice differences in teaching: A cognitive analysis and implications for teacher education'. *Journal of Teacher Education*, 41(1): 36–42.

Randall, M. and Thornton, B. 2001. *Advising and Supporting Teachers*. Cambridge: Cambridge University Press.

Roberts, J. 1998. *Language Teacher Education*. London: Edward Arnold.

Schön, D. 1983. *The Reflective Practitioner*. London: Temple Smith.

Senge, P. 2002. 'Moving forward'. In P. M. Senge, A. Kleiner, C. Roberts, R. Ros, G. Roth and B. Smith (eds), *The Dance of Change: The Challenges of Sustaining Momentum in Learning Organizations*. London: Nicholas Brealey Publishing.

Wallace, M. 1991. *Training Foreign Language Teachers: A Reflective Approach*. Cambridge: Cambridge University Press.

Wallace, M. 1999. 'The reflective model re-visited'. In H. Trappes-Lomax and I. McGrath (eds), *Theory in Language Teacher Education*. London: Longman.

Wideen, M. F., Mayer-Smith, J. A. and Moon, B. J. 1996. 'Knowledge, teacher development and change'. In I. Goodson and A. Hargreaves (eds), *Teachers' Professional Lives*. London: Routledge Falmer, pp. 187–204.

3
Learning to Talk the Talk: Conversation Analysis as a Tool for Induction of Trainee Teachers

Paul Seedhouse

Trainee and newly qualified professionals often experience difficulty in becoming proficient in the professional discourse required in both interaction with clients and interaction in meetings with fellow professionals. Some professions have begun to address the need to induct new professionals into professional discourse, often by offering observation or by showing videos of typical interaction. This chapter argues that many of the complexities and subtleties of professional discourse may not always be evident during observation or videos and that it is often precisely these complexities and subtleties which cause the problems for the newcomer. However, these may sometimes be revealed by fine-grained conversation analysis (CA) of transcripts which may then be combined with video to create a powerful induction tool.

The professional context we consider in this chapter is that of the English language teacher (EL teacher henceforth). En route to becoming a successful EL teacher, there are many skills which a trainee needs to develop. Some of these may be taught on teacher training courses, whilst others may be learnt through trial, error and bitter experience. A particular puzzle for trainee teachers is how it is that experienced teachers manage to create a pedagogical focus, that is, to get students to do what they want, in an apparently effortless manner. When the trainee teachers give instructions, however, the students often don't understand what to do, the target pedagogical focus is not created and confusion results. There is often a mismatch between what the trainees want the students to do and what the students actually do. In this chapter we will try to unravel this puzzle from two angles. We will examine an example of what trainee teachers sometimes do wrong and show how and why the instructions which they give manage to confuse students. We will also look at an example of what experienced teachers typically do right and how they give instructions so that the students are able to carry out the required procedures. In order to do so we will examine in very fine detail transcripts of interaction involving experienced and trainee teachers.

Establishing a pedagogical focus

We will firstly examine how an experienced teacher successfully establishes a pedagogical focus. The data are from a language school in Mexico. The teacher has previously asked the learners to bring a personal possession to the class which is special to them in some way. In EL teaching, the idea of a contrast between a focus on form and accuracy and a focus on meaning and fluency is a common and widely accepted one (e.g. Brumfit 1984; Seedhouse 2004; Widdowson 1990). In this lesson, we will examine how the experienced teacher establishes first of all a pedagogical focus on meaning and fluency and then manages a shift in focus to form and accuracy.

Extract 1

T = Teacher
T: Today's class is going to be about describing objects, and we're going to look at three different types of description. I'm going to write it here on the board, what we'll be doing. (T writes on board) The first type will be 'personal' OK? Objects that have an especial value for you, a personal value. The second type will be catalogue type descriptions.

(British Council 1985, Volume 4: 50)

The lesson starts with a procedural introduction which anticipates that the lesson will involve some kind of change of focus and which provides a link between the two focuses, in that they will both involve description. For the next stage in the establishment of the focus, the teacher asks the learners if they have brought personal belongings along as requested, and elicits from two or three students the nature of their belongings. Then the teacher produces an enormous embroidery, a personal belonging with personal value for her, and tells the learners about it:

Extract 2

> T: Um, this is a nineteenth century, Japanese embroidery, and it was given to me by my great-aunt. My great-aunt, she had a, a funny kind of job really, she was a governess.
>
> (British Council 1985, Volume 4: 51)

This part of the interaction serves multiple functions. It establishes the nature of the focus, in that the teacher is demonstrating what the learners are to do during this episode of the lesson, that is, describe their personal possession and its significance to them. It establishes the nature of the speech exchange system, that is, a monologue addressed to the other participants. The teacher has stated that a purpose of this part of the lesson is for learners to learn a bit about each other, and here the teacher is telling the learners something about herself and thus developing her relationship with the learners. The teacher then rolls up her embroidery and issues further procedural directions for the establishment of the focus.

Extract 3

> T: And what I want you to do is to talk about your things now in the same way as I did about mine, saying what it is and give the history of it. How, why have you got it, and maybe also say why is it important to you. For this (0.3) thank you, thanks, can you put it at the back, right that's great (0.3) em, we're going to work in two groups. so, would you be a group of six here: you two, and you four. Can you get into a little circle (0.4) hang on for a sec (0.3) and you're going to be seven here. Can you move your chairs quietly, so it doesn't make too much noise (8.0) Yes, join this group. OK, it doesn't matter who begins. Whoever wants to can, can start. I'm going to come and sit with each group some of the time but just listen.
>
> (British Council 1985, Volume 4: 51)

So the spatial configuration of the learners is altered in preparation for the main context. How does the teacher ensure that it is in fact a 'meaning and fluency' focus which is established rather than any other? This appears to be accomplished in the following ways: (a) by explicitly modelling the type of talk which is to be produced, which implicitly establishes a context; (b) by giving explicit instructions concerning the nature of the speech exchange system and the topic of the talk; and (c) by focusing on the content of the talk and by not mentioning linguistic accuracy. The teacher states 'I'm going to come and sit with each group some of the time but just listen.' The use of 'but just' implies that the teacher will not be conducting repair of linguistic errors, and hence that the emphasis should be on the expression of personal meanings.

Extract 4

L1 = Learner 1, and so on.

001	L1:	OK. As you see this is a music
002		box, and my mother made it. It's
003		(0.5)
004	L2:	Oh, your mother made it.
005	L1:	Yes, my mother made it. The thing
006		is that when, this is the first
007		thing she did like this, with
008		painting and everything, so
009		nobody, nobody thought that it was
010		going to come out like this.
011		That's the point. That's why this
012		is special because it took her
013		about three weeks to, to make it,
014		and er she, she put a really
015		special interest in that, and
016		tried to, to make it the best
017		that, er she could.

(British Council 1985, Volume 4: 51)

We can see from Extract 4 that the interaction produced by the learners is as expected within a 'meaning and fluency' focus. The learners express personal meanings, and linguistic errors (as in lines 10–14) are ignored. Heritage (1984) notes that 'oh' often occurs as a marker of change of information state, since new information is being exchanged. We can also see that the learners are managing the speech exchange system themselves. Although the teacher modelled a monologue, L2 feels able to self-select and disrupt the monologue (line 3). So the teacher has used multiple methods of ensuring that the correct pedagogical focus is created, and in this case the intended focus has clearly been successfully created.

Managing focus shift

We will now look at an example of how a focus shift is managed by an experienced teacher. Looking at the same lesson, we will see how the previous 'meaning and fluency' focus is shifted to a 'form and accuracy' focus. The teacher brings the previous focus to a close in the following way:

Extract 5

001	T:	OK. Can I stop you now? I know not
002		all of you have finished but we
003		haven't got time for any more so
004		let's get back into two lines
005		again.
006		((LL move chairs))
007		Some really nice objects there.
008		Whose was the oldest? I think
009		Lena's was. When do you think your
010		object is from?
011	L1:	From ( ) it's from near Mexico
012		City and long time ago it was a
013		lake (0.5) now it's a long time ago
014		it's a lake.
015	T:	And how old do you think that is?
016	L1:	Well (.) er (.) I suppose it is
017		200 years old (0.8) I suppose it at
018		least. Probably more, probably
019		more, yes.
020	T:	Yes, maybe even (0.8)
021	L1:	300
022	T:	4 or 500.
023	L1:	Yes. ((LL finish moving chairs))
024	T:	Um, OK well remember that I said
025		the second thing we're going to
026		look at is catalogue type
027		descriptions. Sometimes when we're
028		describing things we need to use a
029		lot of different adjectives and
030		sometimes we're not very sure

031	which order we should put the
032	adjectives in. For example do we
033	say um (0.8) 'A green felt hat' or
034	'A felt green hat'. OK which way
035	round should we put the
036	adjectives? So we're going to
037	take a look today at this chart
038	((T points to chart)) which gives
039	us an idea of how the order of
040	adjectives should go. So first of
041	all we have, where we normally
042	described things, first of all we
043	have age. You don't need to copy
044	it down because I'll give it to
045	you in a minute. We've got age,
047	then size, shape, colour, manner,
048	place, material and use or
049	function. All right, well what
050	we're going to do is I am going to
051	give you a handout and on the top
052	you've got some jumbled sentences.
053	OK. These are just little
054	descriptions but the adjectives
055	are all in the wrong order. I want
056	you to work in pairs to put them
057	into the right order. And these
058	are three, A, B and C, then I'd
059	like you to do D and E, I'd like
060	the pair of you to write a couple
061	more descriptions using lots of
062	adjectives. OK. Does everyone
063	understand? All right Could you
064	give these out – yeah Gracia – can
065	you give those out – pass one
066	along. (LL give out sheets) Look
067	up at the chart, use the chart as
068	much as you need to, to help you
069	get the sentences right.

In line 1, T explicitly marks a transition. 'OK. Can I stop you now?' is uttered with high pitch and high volume. T indicates that there is to be a change in spatial configuration back from small group to whole class in lines 4–5. Now whilst the learners are moving chairs back the teacher engages in some interaction which still has a 'meaning and fluency' focus, from lines 4–23. The topic of the interaction is a personal possession and linguistic errors are not corrected by the teacher. It may at first sight appear confusing that the teacher should initiate some kind of shift and then return to the previous context. However, this temporary return is conducted whilst the learners are moving their chairs and terminates as the learners finish their spatial reconfiguration. Furthermore, this temporary return also involves a positive evaluation and appreciation of the terminated activity ('some really nice objects there').

In line 24, T arranges the focus shift. This starts with 'OK well' (which function as topic disjunction markers) uttered with high pitch and volume and continues (in lines 24–27) with a reference back to the procedural stage at the start of the lesson, where it was indicated that the second phase of the lesson would be concerned with catalogue type descriptions. The teacher develops a form and accuracy focus in the following way. There is a focus on linguistic correctness in the expressed concern for the proper order of adjectives (lines 34–36). The change in focus is symbolised by the presentation of a chart of the correct order of adjectives (lines 36–38). The teacher distributes materials in which the adjectives are in the wrong order, with the instructions that the learners are to put them in the right order (lines 55–57). A focus on form and accuracy and linguistic correctness without regard to personal meanings is thus established. Whereas in the previous 'meaning and fluency' focus the learners supplied the materials (which were personally meaningful and which they had to hold close to themselves), in this EL classroom context the teacher presents the materials in a 'logical', impersonal chart format which is placed at some distance from the learners. This change seems to indicate on a semiotic level to the learners that they no longer have any interactional space to express personal meanings. Finally, the teacher introduces a major new pedagogical focus (lines 55–57) which is incompatible with the focus inherent in the previous context.

Shifts in focus always seem to be distinctly marked in the data in some way by successful, experienced teachers. They may be marked by use of discourse markers, by prosodic features, by changes in the spatial configuration of the participants, by meta-discoursal comments which indicate that a shift is occurring and by semiotic means. This area is further explored in Seedhouse (1996).

An example of failure in the establishment of a focus

As Levinson (1983: 319) points out, a key source of verification that an interactional organisation is actually oriented to by the participants rather than being an artefact of analysis is what happens when a hitch occurs in the organisation. It will now be argued, by examining a 'deviant' case of an extract from an EL lesson taught by an inexperienced teacher who is not yet fully competent in establishing a focus, that the ability to create and manage a pedagogical focus is not something automatic and given, but a skill or competence which is learned, and that an important part of being a competent teacher is the ability to create, manage and shift the pedagogical focus. The data are from Seedhouse (1996); it is an English lesson in a British language school, and the teacher is a trainee. It should be noted that the lesson as a whole is fairly successful and that the trainee is having a little local difficulty in this episode of the lesson.

Extract 6.1

001	L1:	I was drive (0.5) drive drive
002		driving a car?
003	T:	I was driving a car?
004	L1:	eh when (0.5) you:: (1.0) eh (1.0)
005		um (0.5) drink a=
006	T:	=when you=
007	L1:	=when you drank drank a: a orange
008	T:	when you drank an orange. OK you
009		were driving the car (0.5) when
010		you drank an orange.
011	L1:	yes
012	T:	(0.5) OK?
013	L1:	haha
014	T:	huhu strange but it's OK correct
015		OK right (0.5) this time let's
016		just think ((looks at textbook))
017		about these children of courage

018		we've got Mark Tinker? (0.5)
019		who's aged 12 comes from London
020		(0.5) Jackie Martin 14 comes from
021		Manchester (0.5) and Daniel Clay
022		who's 13 and comes from Newcastle.
023		(0.5) right can you see the
024		pictures? (0.5) can you see them
025		Malta?
026	LL:	(xxxxx)
027	T:	right children of courage what do
028		you think (0.5) children of
029		courage will do? (2.0) what do
030		children of courage do. (1.0) or
031		what did they do rather what did
032		they do? (2.0) what does courage
033		mean? what's this idea if I am
034		courageous (2.0) how would you
035		describe me? (2.5)

(Seedhouse 1996: 306)

A shift of focus takes place in line 15. Prior to line 15 the participants were operating in a form and accuracy focus, in which learners had to construct sentences which combined the past continuous and the past simple. We can see from the teacher's comment in line 14 that the fact that the learner has produced a bizarre sentence is unimportant, since the focus is on the production of a string of formally accurate linguistic forms without regard to 'meaning'.

In line 15 the teacher attempts to shift the focus from 'form and accuracy' to a text-based focus. The students have a textbook open in front of them; the text is entitled 'Children of Courage' and has various photographs of children with stories of their courageous acts. The teacher attempts to signal the change to a text-based focus by shifting her gaze towards the textbook simultaneously with starting to read information from the textbook concerning the characters. There is also use of shift markers together with very slightly raised pitch and volume. However, what is noticeable in the video and transcript is that the shift of focus is not marked very strongly. Also, there is no meta-discoursal explanation about the shift or the nature of the new focus. This is in contrast to the heavy emphasis placed on these by the experienced teacher.

The teacher therefore appears to have shifted to a text-based focus, but the precise pedagogical focus is unclear, as we shall see. At first (lines 27–32) the teacher appears to want the learners to predict the content of the story ('what do you think children of courage will do?'), and then to describe the content of the story which they have not yet read ('what did they do?'). Then the teacher tries to elicit the meaning of a single lexical item (lines 32–33), and then asks the learners to supply a description of herself (lines 34–35). So although the learners can be fairly clear that there is now a text-based focus in that they are apparently being required to look at the text and supply an answer from the text, they have been given four contradictory sets of pedagogical focuses by the teacher.

Extract 6.2

036	L2:	I describe one person?
037	T:	yes well anybody if if you (0.5)
038		were (0.5)one of these children of
039		courage (6.0)
040	L3:	don't understand
041	T:	you don't understand. OK people of
042		courage. what would they have
043		done? what do you think they do?
044		(0.5)
045	L4:	he is on holiday?

(Seedhouse 1996: 306)

In line 36, L2 has latched onto the teacher's last instruction (lines 34–35) and tries to clarify whether the required aim is to describe the characters in the text. T's utterance in lines 37–39 does nothing to clarify the issue. L3 also indicates non-comprehension in line 40, but rather than clarifying which of the four sets of aims that have already been introduced the learners should focus on, the teacher actually takes a previous question (from lines 29–30), changes the subject from 'children' to 'people', and changes the tense of the question twice (into the rather difficult conditional perfect and conditional forms), thus confusing the learners further. L4 assumes that the required aim is to describe what the characters in the text are doing and provides an answer from the textbook in line 45. At this stage it is clear that the students have no idea what the intended pedagogical focus is.

Extract 6.3

046	T:	they're on holiday? no but to be
047		courageous do you understand the
048		word courageous? courageous? (0.5)
047	L:	no I don't
048	T:	no? courageous (4.0) courageous
049		(2.0) what would you have done?
050		(2.0) no?
051	L:	no
052	T:	no idea (1.0) OK for example (1.5)
053		somebody is lying in the road
054		what do I do? (4.0) no? or (1.0)
055		I'll carry on the action (1.0), I
056		come in I see this person on the
057		road and I run to them (1.0) I see
058		if they are alive they're not
059		breathing (1.0) so: I turn them
060		over and I give them (0.5) the
061		kiss of life for example (1.0)
062		give them the kiss of life and
063		they begin to breath (3.0) then I
064		go to the telephone box and I ring
065		an ambulance (1.0) I come back to
066		my casualty and I have saved: his
067		life (2.0) right? so me, I am
068		courageous yeah? I'm courageous
069		I'm brave I've done something I've
070		helped this person (1.5) yeah? do
071		you understand?
072	LL:	yeah
073	T:	Ok so if there has been an
074		accident, for example (1.0) any
075		accident and somebody comes to
076		help and they find someone in a
077		difficult situation and they are
078		practically dying or er maybe they
079		are erm in a dangerous place, for
080		example on a cliff on a mountain
081		in the sea or something like that

082		in a lake. they fell in (1.0) I
083		come along and they need help
084		because they can't get out, right?
085		and I help them (1.0) the person
086		who helps them would be courageous
087		(1.0) do you understand? brave so
088		all these children here it says
089		children of courage OK? they had
090		courage they helped they saw a
091		danger and they helped the person
092		(1.0) OK?
093	LL:	hhm

(Seedhouse 1996: 306)

In line 46 we can see that it has become clear to the teacher that the students do not understand the key word 'courage'. She therefore narrows the pedagogical focus down to the meaning of the lexical item 'courageous' and provides two long explanations. The learners indicate that they understand the meaning of the word, although this is not verified.

We can conclude from the above data that, if a change in focus is to be undertaken, it is essential for the teacher to make the nature of the pedagogical focus as clear and as explicit as possible. Presenting multiple pedagogical focuses simultaneously is likely to confuse learners.

How do experienced teachers deal with failure in the establishment of a focus?

Now if we contrast the way that experienced teachers deal with communication problems, we may be able to draw certain preliminary conclusions. With experienced teachers, few examples of miscommunication in relation to procedural matters occur in the data. Experienced teachers do not generally issue elaborate procedural instructions in order to set up contexts; they tend to be simple, clear and focused. In Extract 7 below, we see how an experienced teacher deals with miscommunication concerning procedural matters.

Extract 7

001	T:	I've got a sofa a sofa say after
002		me I've got a sofa

003	LL:	I've got a sofa
004	T:	very good and now I need Kjartan
005		and Elge (0.5) can you come up to
006		me please (1.0) and can you give
007		each one a sheet?
008	L:	sheet?
009	T:	sheet of paper (LL hand out sheets)
010	T:	now again (2.5) listen to me (6.0)
011		I've got a lamp
012	LL:	⌜I've got a lamp
013	T:	⌊what
014	T:	don't repeat now, don't say after
015		me now, I I say it and you and you
016		just listen. I've got a lamp. What
017		have you got? (2.0) raise your
018		hands. What have you got Eirik?
019	L1:	e:r=
020	T:	=can you say=
021	L1:	=I've got a book.
022	T:	right, fine. I've got a telephone
023		what have you got? (3.5) Trygve.
024	L2:	I've got a hammer.

(Seedhouse 1996: 314)

In the above extract there is a small change in procedure and focus. Line 2 is the final line of a procedure in which learners repeat what the teacher says. From line 10, however, the procedure is that each student has a different object. The teacher says 'I have a lamp. What have you got?' to an individual student, who replies 'I've got a...' according to which object the student has. This is indicated by movement around the classroom and the handing out of sheets. In line 12 some of the learners try to repeat what the teacher says, that is, they are continuing the procedure and speech exchange system from the previous episode. When the learners show signs of having misunderstood the procedure, the experienced teacher: (a) makes clear in line 14 what the learners have done wrong; (b) narrows the focus down in lines 15 and 16 by repeating and clarifying the procedural instructions in very simple terms. When the trainee teachers are confronted with miscommunication concerning procedural instructions, however, the strategy they adopt is virtually the opposite of that of the experienced teacher. They move away from the procedural instructions already given and enlarge or dilute the focus considerably by issuing multiple differing and even mutually contradictory pedagogical focuses.

Conclusions

We can see from these different cases that a pedagogical focus is actively constructed and maintained by experienced teachers. Creating and shifting a focus is a skill which is acquired through experience. Without careful management, there can be confusion as to what the focus is at any given time. The unsuccessful extract was taught by an inexperienced trainee teacher and it is possible to draw certain conclusions from it. It is easy to confuse learners with respect to classroom procedures and as to which focus is in operation at a particular time. It is best to state explicitly what the pedagogical focus is, and it is best to introduce one pedagogical focus at a time, otherwise learners may become confused. It is best for procedural instructions to be as full and explicit as possible whilst presenting a single, undiluted focus. A similar point is also made by Johnson:

> Explicit directions and concrete explanations can help second language students recognize the implicit norms that regulate how they are expected to act and interact in classroom events. Without such explicitness, second language students can become confused about what is expected of them, or how they should participate.
>
> (1995: 163)

It is necessary to look in the micro-detail of the interaction to establish how and why classroom procedures succeed or fail. It is unlikely that this level of detail will be evident to trainees in videos alone, whilst transcripts alone are not able to contextualise the interaction vividly enough. I have shown the video of the trainee teacher failing to establish the pedagogical focus a number of times in seminars and classes. After watching the video but without having seen the transcripts, participants are normally able to identify the problem, that is, that the teacher has failed to establish a focus. However, participants are generally unable to establish the reason for this failure. It is only when the participants read the transcript that it becomes evident to them that there are clear reasons for this failure and that they are able to identify these.

It is suggested, in conclusion, that fine-grained CA analysis of transcripts may be combined with video to create a powerful induction tool into professional discourse for trainee or newly qualified EL teachers. A general framework for this process might look as follows:

1. Make videos and transcripts of both experienced and inexperienced EL teachers in a variety of typical professional situations with both other professionals and students.
2. Identify in the fine detail of the interaction those interactional issues which may lead to a more or less successful conclusion of the interaction.
3. Identify in the fine detail of the interaction those key interactional devices which are used by experienced professionals and analyse how they use them. An example in this chapter is the establishment of a pedagogical focus by an experienced teacher. An example from another professional context is Drew's (1992) explication of a device used by lawyers for producing inconsistency in, and damaging implications for, a witness's evidence during cross-examination in a courtroom trial.
4. Disseminate findings to trainee and new professionals using video combined with transcripts.

Individual teachers who are not in a teacher training context could also employ CA as a tool for their own professional development. This would involve teachers video recording their own lessons, or working jointly with a colleague on recording each other. The teachers would then transcribe and analyse the micro-detail of their lessons. Areas which might be focused on in analysis are:

- Sequences in which trouble of some kind occurs
- Sequences which went particularly well and in which successful learning was thought to have taken place
- Lesson transition sequences and how the learners oriented to these
- Sequences in which the teacher produces instructions or explanations
- In action research, the teacher might record a 'default' lesson, then introduce an innovation into the teaching context which is then recorded and the two lessons compared
- What actually happens in pairwork and groupwork?

References

British Council. 1985. *Teaching and Learning in Focus: Edited Lessons* (4 Volumes). London: British Council.

Brumfit, C. 1984. *Communicative Methodology in Language Teaching.* Cambridge: Cambridge University Press.

Drew, P. 1992. 'Contested evidence in courtroom cross-examination: The case of a trial for rape'. In P. Drew and J. Heritage (eds), *Talk at Work: Interaction in Institutional Settings.* Cambridge: Cambridge University Press, pp. 470–520.

Heritage, J. 1984. 'A change-of-state token and aspects of its sequential placement'. In J. M. Atkinson and J. Heritage (eds), *Structures of Social Action.* Cambridge: Cambridge University Press, pp. 299–345.

Johnson, K. 1995. *Understanding Communication in Second Language Classrooms.* Cambridge: Cambridge University Press.

Levinson, S. 1983. *Pragmatics.* Cambridge: Cambridge University Press.

Seedhouse, P. 1996. 'Learning talk: A study of the interactional organisation of the L2 classroom from a CA institutional discourse perspective'. Unpublished PhD Thesis. University of York.

Seedhouse, P. 2004. *The Interactional Architecture of the Language Classroom: A Conversation Analysis Perspective.* Malden, MA: Blackwell.

Widdowson, H. 1990. *Aspects of Language Teaching.* Oxford: Oxford University Press.

Reflections on Starting Out

Wei Liu

Being absolutely at the stage of 'Starting Out' in my ESOL career, I have really benefited from reading the chapters written by Copland, Kurtoglu-Hooton and Seedhouse.

English is not my native tongue. I have been learning English for 25 years. During this time my passion for the English language has gradually developed into a commitment to promoting intercultural communication, fuelling my linguistic and cultural study.

I come from a teacher-centred culture with a passive learning style. As a matter of fact, my bachelor's degree in Engineering didn't provide any direct guidance for my two years' part-time adult English teaching experience. Embarking on MA TESOL Studies in Aston University has introduced me to a very interactive student–centred learning style. I have been able to deepen theoretical and methodological studies and put theories into practice to become more competent in teaching. But a few teaching practice sessions are not enough to start a real career path. I am therefore looking for more practice sessions and am also employing the techniques I have learned in teaching Mandarin Chinese.

As Copland notes, when I first started teaching practice sessions, I was confused to see the judgement criteria on the feedback form and also admired the sample lesson-plan produced by previous trainee teachers. On reading Copland's chapter, I felt I was in tune with being a trainee in that I am eager to get guidance on how to be a better teacher. In this respect, my peers and I were greatly supported and guided by our tutor, who gave so many practical instructions. She taught us everything from general rules of physical positioning in a classroom (facing all the students), to technical improvement on teaching pace and methodologies. We all put these suggestions into practice straight away and felt we had benefited greatly from these improvements. Copland's discussion in the section on 'roles and relationships' reconfirms my belief that a trainee should put development prior to assessment. The importance of face-saving, especially for those in the learning process, is widely recognised in Chinese culture, and so this technique is one that I would like to use myself. I learned the importance of prompting trainees to comment on the performance themselves and identify their own errors. In one of my practice sessions, I called out learners' names to correct their mistakes made in the class. This was a good lesson to learn.

Kurtoglu-Hooton's chapter on confirmatory and corrective feedback is highly instructive for both tutors and trainees. Coming from a learning style dominated by a corrective approach, I am very familiar with the difference made by these two different forms. There must be a positive way to describe the corrections, that is, substituting 'please don't do that' with 'please do this'. I might also suggest using a 'Confirmatory-Corrective-Confirmatory' approach to enhance this proposal. The vignette of Jack's grouping technique inspired me to expand this kind of particular technique confidently and repeatedly. I recalled that in my own MA TESOL course everyone enjoyed going to one particular module because not only did it have a relevant and engaging teaching content, but in addition the lecturer always gave interesting warm-up activities for us to start with.

The results of Seedhouse's research on the importance of introducing one pedagogical focus at a time support my strategy of setting realistic plans and objectives in the teaching practice sessions. There are an overwhelming number of methodologies that could be adopted and experimented with in teaching practice and so trainees are under great pressure to demonstrate the skills and techniques that are needed to be assessed. It is unrealistic for trainee teachers to achieve all the intellectual development as well as the practical development during a short training programme or a degree course. What I have found very challenging in my own learning is to shift smoothly from concentrated practice on one skill (e.g. listening, speaking, reading and writing) to another within the limited time scale and provide a seamless connection. Seedhouse's case study of managing a shift in focus from meaning and fluency to form and accuracy really helped me in this aspect.

The chapters written by Copland, Kurtoglu-Hooton and Seedhouse about trainee teachers and tutors highlight the aspect of 'doing'. I would also like to discuss the importance of 'being' a trainee teacher which has a great impact on future learners who will learn from our examples. Future learners will notice the discrepancies between what I practice and what I preach, that is, I should show myself as an example of what I try to encourage and expect the learners to do.

When a teacher starts to teach, it is not important whether he or she knows a great deal about the theories of teaching. What matters are attitude, desire and dedication. Teaching is itself a science and an art which can be gradually learned through study and experience. The attitude of respect for students and the recognition of their efforts determine the tone of teaching. It is a confirmatory attitude which naturally leads to a confirmatory feedback approach. I find this quotation from the Persian philosopher Baha' u' llah very inspiring:

> Should any one among you be incapable of grasping a certain truth, or be striving to comprehend it, show forth, when conversing with him, a spirit of extreme kindness and good-will. Help him to see and recognise the truth, without esteeming yourself to be, in the least, superior to him, or to be possessed of greater endowments.
>
> (trans. Shoghi Effendi, 'Gleanings from the writings of Baha'U'llah'; IV: 6)

I try to ask myself each time I teach, the reason why I am teaching this lesson. This is not just a process of learning together and passing on skills, theories and methodologies. The most important aspect of any lesson is the experience itself and the impact the teaching will have on the lives of the learners.

I have learned that when I get stuck in class, there is no need to panic. In a lot of situations the unexpected will happen to challenge trainee teachers' knowledge, lesson preparation and classroom management skills. Any factor could undermine the confidence of a new teacher, especially when a trainee who is not a native English speaker is being questioned about an area they are not familiar with. The best strategy is either to pause and revisit it at another stage or occasion, or turn the situation into a more planned pedagogical focus.

I would encourage trainee teachers who are at this starting out stage to have the confidence to make mistakes and learn positively from them. We must appreciate the experience of making our first mistakes and understand how this will strengthen us as teachers. 'Failure is the mother of success', as the Chinese saying testifies. Always prepare a contingency plan and get ready to improvise. I illustrate this idea every time I forget to bring a board pen with me! Most importantly, rehearse before the actual lesson. The more thoroughly the lesson is prepared, the more successful it will be. Furthermore, we must be careful not to throw ourselves in at the deep end right at the beginning. There is a great demand for English language learning all over the world. There are vacancies either to teach a mixed group with various levels, or to start as a floater wherever there is a gap that needs to be filled. Before plunging into a full-time job with increasing pressure, a novice teacher should have a realistic and steadily developing plan. This step will make a great impact on teaching motivation and the sustainability of one's career path.

I would also like to contribute a few ideas for the tutors, educators and training providers to consider.

1   A tutor as a mystery shopper?

Both tutors and trainees hope that the teaching practice sessions will be conducted in an atmosphere that is as relaxed as possible. Technology permitting, a teaching practice without a tutor on the spot would be desirable to release the pressure on the trainee. This could be achieved by employing a 'mystery shopper' who is a third party with all the competence and skills to observe the session and provide the assessment. This method could be reinforced by the use of recorded video and audio data for self–assessment and tutor assessment purposes.

2   Allow trainees to practice and progress in their own time

The tutors are not there just to identify the trainee's mistakes, but also have the responsibility to provide tailored suggestions for their correction and progression. It is obvious that it is not possible to correct all the mistakes a trainee might make during training. The identification of mistakes lays the most significant foundation. Therefore tailored suggestions and action plans might be the best solution for this situation. This would leave the trainee teacher able to progress at a more realistic pace.

3   A mentor for guidance

Resources permitting, it would be ideal for a trainee teacher to be paired with a mentor who is at the *passing on knowledge* stage. It could be beneficial for both parties to exchange enthusiasm and new vision for expertise and experience. More importantly this may improve the process of acquiring skills, as mentioned by Seedhouse which were; '...learnt through trial, error and bitter experience'.

As a trainee, it is my goal to become a motivating, respectful and supportive (MRS) teacher. I hope to blaze through the challenges which lie ahead and aid learners to see their half-full cup of water gradually fill to the very top.

Part II

Becoming Experienced

Paul

Paul has seen those films where time suddenly slows down and allows the hero to accomplish impossible tasks at normal speed in a slow-motion world, and it occurred to him the other day that the same thing has happened to him. When he started teaching his lessons disappeared in a flash, a panic of activity that left his head buzzing, but now he seems to have space to think. This has opened up his teaching because it allows him to make on-the-spot decisions in the classroom and explore aspects of his practice. He has the confidence now to think about the sort of teacher he wants to be.

Paul has reached the 'stabilisation' phase in his teaching at the end of 'a progression toward mastery or expertise, achieved some time in the fourth year of teaching or beyond' (Liston, Whitcomb and Borko 2006: 352). He can now rethink some of his key ideas and practices so that he is in a position to respond to new situations (Hammerness, Darling-Hammond and Bransford 2005). This is a time of professional consolidation, when teachers benefit from the confidence that comes from the achievement of technical mastery and integration into their peer group. They are free to explore the dimensions and possibilities of their practice and of their professional identity (for an overview of research on teacher identity, see Beijaard, Meijer and Verloop 2004).

Garton's chapter explores an aspect of teacher identity that is particularly pertinent at this stage. The importance of teacher beliefs has long been recognised, and it is generally accepted that there is a strong connection between these and classroom practices, but evidence of direct connections between belief and action has proved tantalisingly elusive. What is unusual about Garton's chapter is that she is able to identify

two very different and fundamentally distinct belief systems manifested by two teachers involved in very similar work, and to show how
their approaches to teaching are also distinctly different. Her careful
analysis of interaction patterns conveys a strong flavour of their classrooms and the efficacy of their very different approaches.

Garton's research is not designed to categorise teachers, but to demonstrate the importance of beliefs and to underline the need to understand the springs of our action. Her chapter concludes with helpful
advice about how teachers can draw on their awareness of their beliefs-
in-action as a basis for the productive exploration of their practice.

Howard's chapter also explores two very different sorts of lesson, but
the springs this time are external rather than internal. Sooner or later
most teachers will find themselves in a situation where their teaching
will be observed, often as part of an appraisal system over which they
have no control. Teachers know that this is an unnatural situation and
that the resulting lessons are atypical, but what actually happens in
them has remained something of a mystery as far as researchers are
concerned.

Howard opens a window on this secret world and shows how different these 'model lessons' really are from their quotidian counterparts.
More importantly, she demonstrates that their nature is to a large extent
determined by the views of the observer about what makes a good lesson, so that they are to all intents and purposes command performances, designed not to reflect the teacher's normal approach but to meet
the expectations of an external audience. These findings may give
observers cause to reflect on how they go about their work, while her
recommendations may be of help to teachers who have to prepare for
these challenging performances.

A more pleasing aspect of an established teacher's work is the understanding that comes from a growing awareness of students' capabilities
and the confidence to explore new ways in which classroom work might
be linked to language development outside the institution. The rapid
expansion of global tourism means that more students now have the
chance to interact with English speakers, but these encounters are often
frustratingly short and unsatisfying. Morris-Adams's chapter, however,
demonstrates the potential of such exchanges.

In choosing to focus on topic management, Morris-Adams explores a
feature of talk that is essential for extended conversation and yet usually absent from classroom exchanges. In examining the strategies used
by non-native speakers in encounters with native speakers, she reveals
a range of competences that can be drawn on to exploit the potential of

such talk and makes a strong case for allowing space in lessons to encourage the development of relevant skills and competences.

As Maneerat Tarnpichprasert points out in her commentary, Morris-Adams's work might usefully be extended to encounters between speakers whose first language is not English, while teachers should encourage students to seize opportunities for exchanges with native speakers. Her commentary also reveals the concerns that teachers feel about the challenges of adapting their approach to the needs of very different groups of learners and offers an interesting suggestion about how teacher observation might concentrate more on aspects of change and development than a single, 'model' performance.

Perhaps Tarnpichprasert's most telling reflection, however, relates to the period that leads up to stabilisation, during which she had doubts about her choice of teaching as a profession. The sense of happiness to which she refers may have played an important part in leading her to the point where stabilisation and career acceptance go together because, as Liston and colleagues note (2006: 354), the 'emotional texture of the beginning years' can influence whether teachers stay in the profession. As we shall see in the next part of this collection, those who stay the course find their options opening out.

References

Beijaard, D., Meijer, P. C. and Verloop, N. 2004. 'Reconsidering research on teachers' professional identity'. *Teaching and Teacher Education*, 20: 107–28.

Hammerness, K., Darling-Hammond, L. and Bransford, J. 2005. 'How teachers learn and develop'. In L. Darling-Hammond and J. Bransford (eds), *Preparing Teachers for a Changing World: What Teachers Should Learn and Be Able to Do*. San Francisco, CA: Jossey-Bass, pp. 358–89.

Liston, D., Whitcomb, J. and Borko, H. 2006. 'Too little or too much: Teacher preparation and the first years of teaching'. *Journal of Teacher Education*, 57(4): 351–8.

4
Teacher Beliefs and Interaction in the Language Classroom

Sue Garton

Introduction

> It's interesting how very similar lesson plans carried out by two teachers with totally different beliefs actually lead to formally similar lessons but with very different class atmospheres

The above extract comes from a diary I kept during a three-year case study of two EFL teachers working in Italy, Charlotte and Linda. The diary extract would seem to indicate that teachers can give lessons that are very similar in their organisation and yet are somehow very different in what we might call the social affective climate of the classroom (Legukte and Thomas 1991). In this chapter we will look first of all at the beliefs that two experienced teachers hold about teaching and learning and then investigate whether their beliefs can be seen to influence the interaction patterns that they set up, and thereby go some way towards accounting for the differences in their classrooms.

Background

As an area of research, teachers' beliefs is relatively recent in TESOL, with most studies being carried out only in the last 15 years or so (Borg 2003). However, it is clearly central to understanding what teachers do and why because, as Johnson points out, teachers 'beliefs have an effect on what teachers do in the classroom insofar as beliefs affect perception and judgement' (1994: 439). Therefore, understanding teacher beliefs is fundamental to understanding their classroom behaviours, including the ways in which they interact in the classroom.

The importance of greater understanding of why teachers interact the way they do is given by the realisation that the type of interaction taking place in the classroom plays a fundamental role in successful language learning. Classroom discourse is a form of institutional talk (Drew and Heritage 1992) and as such has its own characteristics, as a result of which interaction patterns may be highly constrained, reflecting the asymmetrical role relationship between teachers and learners and where the teacher generally has responsibility for organising the interaction that takes place there.

If we can understand why individual teachers may favour certain interaction patterns over others, then it would represent an important step forward for teacher education and development too, as it is now widely recognised that:

> one of the useful roles for teacher education may be to find ways in which teachers can articulate and reflect upon what beliefs motivate the interactions they set in train in the classroom.
>
> (Burns 1992: 64)

The study

The data presented below are part of a wider study, the aim of which was to investigate whether teachers working in TESOL could be seen to have coherent personal belief systems which inform their approach to teaching and learning. Data were collected through a series of semi-structured interviews and classroom observations, all of which were audio-recorded.

In common with most recent studies in this field, a qualitative, interpretivist approach was taken because the aim was to investigate teachers' own accounts of teaching and learning by giving individual teachers a voice and focusing on their perceptions, language, actions, thoughts and feelings (Johnson 1994: 441). As Richards (2003: 9) notes, a qualitative approach to research is 'a person-centred enterprise' which involves trying to improve our understanding of people in their natural context.

The teachers

This chapter focuses on two teachers with very different beliefs, Charlotte and Linda, both of whom worked in Italy.

At the time of the study, Charlotte had been teaching English for eight years. She worked freelance for various private schools, organisations and companies, including a university language centre where Linda also worked. During the time we were working on this project, she got a full-time job in a university teaching EAP to undergraduates and consequently gave up most of her freelance work.

Linda had been teaching EFL for nearly 17 years. She worked full-time at a university, teaching EAP to undergraduates. She also did a lot of extra work, including evening courses at the university language centre.

Charlotte and Linda have very different beliefs about teaching and learning, and the following sections, which are necessarily selective, will use some extracts from the interviews carried out with them to illustrate these differences.

Charlotte and the personal side of teaching

In all her talk about teaching and learning, Charlotte focuses very much on what might be called the personal, affective side of teaching. She places a definite emphasis on people, underlining relationships and contact. She believes that it is important for teachers to be genuinely friendly and have an ability to establish rapport with learners.

When asked about the qualities of a good teacher, Charlotte[1] immediately lists the personal qualities of the teacher when she says:

> well I suppose being somebody who's got a good relationship with the students and sensitive to students I mean thinking back to sort of school negative teachers sort of sarcastic sort of humiliating not being interested in the students

This is very closely linked to how she sees her own strengths:

> right strengths I think I'm fairly receptive to students I am sort of I like the students I like people so I'm interested in them which I'm sure helps I think that's probably a strength... and I'm very patient with the students

For Charlotte, teaching focuses on enjoyment and interest, on creating a positive affective environment in the classroom. This can also be seen in her belief about what constitutes a successful lesson:

> well that students are engaged in what they're doing that's number one that they're sort of awake ((*laughter*)) awake and engaged yeah

Moreover, Charlotte's classroom decisions tend to be based on her perception of learners' wants and interests, as can be seen when she is asked what sort of activities she prefers:

> I don't know perhaps more sort of fun things more lively things...
> probably because I think that if they're enjoying the lesson ((*pause*))
> I'm sure it's not really good reason ((*laughter*)) they're enjoying it but
> it's ok cause they've enjoyed it even if they didn't ((*laughing*)) learn
> anything they've had a good time

And again in how she talks about choosing a textbook

> well something that will interest that the students will like topics
> that the students will like

As a result of her concern for what learners want, Charlotte believes in giving her learners space. This is clear from what she believes is her role in the classroom. Initially she lists her roles as monitoring, direct-ing, supporting and source of information. When asked which of these she prefers, she replies:

> probably sort of monitoring and a second not a secondary role but
> when the students are doing something when they're the sort of cen-
> tre of the activity...probably just because they're actually sort of
> doing it and they're involved you know they're speaking and it's not
> really like teaching a subject where you're giving them knowledge.

When talking about the roles that learners should take in the class-room, Charlotte believes it is important for them to help each other as, in some circumstances, the learners are better able to support each other than is the teacher:

> (I think) it depends what they're doing but yeah as we said before
> participating or also sort of supportive between themselves among
> themselves so sort of helping each other (working group work) it
> comes out sort of through group work or if you sort of have atmos-
> phere that they don't feel embarrassed to make a mistake you do get
> a sort of classroom sort of culture that builds up sort of in-jokes

Charlotte makes a conscious attempt to involve the learners directly in their own learning by seeking out their preferences and being sensitive to her learners' felt needs, adapting to these if necessary. An example of this can be seen in her policy on the use of the L1 in the classroom. Although she believes that the learners should be strongly encouraged to use English and she hardly ever uses Italian herself, she is willing to compromise in the interests of creating a positive environment:

> because I mean sometimes you can see it's stronger than them they just have to say it in Italian they just have to because they are curious if they're going round asking questions I mean they do make an effort to try and say in English but then you know curiosity's killing them they've just got to especially Mirco ((*laughs*)) (he's so flipping curious) just has to say it...I think students sometimes do need to sort of let off steam and I think if you're too strict you'll probably find that they feel even more frustrated and speak less

Charlotte is clearly aware that there are institutional roles and expectations in the classroom, but she does not emphasise these roles and indeed appears to downplay them in favour of establishing a good working relationship with her learners. For her, the teacher takes mainly a guiding and supportive role and it is her job to create a positive affective environment which creates the best conditions for learning.

Linda and the teaching and learning process

In contrast to Charlotte, Linda focuses on the teacher's professionalism and knowledge:

> the qualities of a good teacher is the teacher has to be credible to the students...she has to seem to know what she's talking about ((*pause*)) she has to be fairly well-organised in that she has to you know have the right material get there in time smile at the students as though she's expecting them to participate and do something well

Her beliefs are characterised by a focus on the learning process, which is seen as central to teaching and learning. She tends to be concerned with her own preparedness and competence as teacher, as well as with the objective needs of learners. The emphasis is on her knowledge and professionalism and on her ability to give the learners the input they need in order to learn.

For Linda, therefore, her strength is based on her knowledge of the subject:

well I try to know what I'm talking about even if maybe I don't always explain it very well

She has strong beliefs as to what are and what are not appropriate roles for teachers and learners in the classroom as can be seen by the fact that she frequently refers to 'what they're there for' or 'it's my job'. What emerges here is the idea of a relationship based on roles with certain expectations on both sides as to what constitutes acceptable behaviour according to the role that each party has, learner or teacher. This relationship is fundamental to the learning process and therefore to learner progress, which is central in Linda's belief system. When she is asked how she would define effective teaching; she is categorical:

well the only possible basis is what the students learn

When asked what constitutes a successful lesson, the reply is:

one where the students go home happy thinking that they've learnt something

Linda believes that her role is to know her subject and transmit it, as well as be friendly. When talking about her role she says:

well I suppose you have to be sort of is it called authority when you correct them on their English you know sort of say this is correct English this isn't correct English or standard or non-standard

She also believes that this is what her learners expect of her:

they expect you to have authority in the sense that they expect you to know what you're talking about you know to sort of have complete mastery of the subject and able to explain

Closely linked to this are Linda's beliefs concerning the roles that learners should take in the classroom, about which she says:

I mean I think it's probably like a sort of personal thing but I don't like it if I think that they're not listening to me while I'm talking and that's one reason why I try to limit teacher class cause obviously the more you do the more of it goes over their heads sort of thing

When asked if she is happy for her learners to take the initiative, she replies:

> I think it's a good thing obviously you don't want to derail the lesson for too long I mean if you've got a set plan which you normally have however vague you want to kind of get there...I don't like them to sort of think that they can take it over and do what they like because well you know I mean that's not their role...they're there for you to give them the indications on what they should be doing

In the classroom, Linda's decisions are based on what she believes is useful and interesting for the learners. At one point Linda says that she does not like doing teacher-class activities. When asked why, she says:

> well no I mean like for speaking I mean I don't think it's I mean it's obviously not very very a very good use of time to have 20 people listening while one of them stumbles over something that you've just tried to explain ((*laughs*)) you know it's not quite apart from the fact that they get bored and start to chatter it's just not very good use of their time really

Linda has strong ideas as to what constitutes acceptable roles in the classroom and these roles are defined institutionally and are seen as functional to ensuring that the learning process moves forward.

Although necessarily brief, it is clear from the above discussion that both Charlotte and Linda share a concern for good relationships in the classroom and for the fact that their learners should learn. However, their beliefs about how this is to achieved can be seen to differ in significant ways.

Charlotte's route to learning is the creation of a positive affective environment in the classroom, where learners are interested, engaged and enjoying themselves. This is the key to motivation and hence learning. In other words, the teacher creates the right conditions and the learner learns.

Linda, on the other hand, believes that the route to learning is via a well-prepared, competent and professional teacher who understands her learners' needs and is able to address them, thereby striving to ensure that the learning process is constantly moving forward.

Let us now turn to the interaction patterns that are found in Charlotte's and Linda's classroom to see the extent to which these are consistent with the beliefs of the two teachers.

Beliefs and interaction

In order to see the effect that beliefs may have on interaction, extracts are presented from two classroom episodes, one for each teacher. These particular episodes were chosen because they are very similar in many ways, which makes them more comparable.

Both lessons are part of general English evening courses organised by the university language centre in Italy. The learners are adults and the majority are university students from different faculties though there are some local people from the town. The lessons last 90 minutes. Charlotte's group is elementary level whereas Linda's group is intermediate.

In both cases the learners did an activity in which they were asked to describe pictures and which was a direct follow-on from the activity that the learners had just completed. Charlotte's group had just finished an exercise where they had to match phrases describing daily routines with pictures about a textbook character's day. They then had to use the phrases to describe the pictures orally. Linda's class had just completed an exercise where they had to experiment with different ways of discovering grammar rules (inductive and deductive). Then they had to apply the rules they had learnt for *so* and *such* to describe pictures. In both cases, the teachers used the technique of doing an example teacher-to-class before the learners were asked to continue the exercise in pairs.

The interaction patterns

It is notable that, overall, Linda's interaction appears to be somehow 'smoother', following a clearer Initiation–Response–Feedback (IRF) pattern (Sinclair and Coulthard 1975). In contrast, Charlotte's IRF sequences are rarely a straightforward three-part exchange and there tends to be much negotiation before arriving at the final F-slot.

These general observations are already consistent with the beliefs of the two teachers. Linda's concern with the learning outcomes and learner progress can be seen in the way that she tries to ensure the learning process is constantly moving forward. Charlotte, on the other hand, is happy to allow for a more 'chaotic' progression by which she can create an affective climate where learners are engaged and not afraid to take risks.

The form that the teachers' I and F slots take is also fundamentally different, as the following examples illustrate clearly.

Initiation

It is immediately clear how the way that the two teachers elicit answers from the learners in order to complete the descriptions of the pictures is very different:

Extract 1

L = Linda; S = Silvia

001	L:	right then let's have a look
002		we'll do this one together and
003		this one together and then I'll
004		let you do this one and this one
005		by yourselves ok so this one ooh
006		look what's happening ((*holds up*
007		*book*)) what vocabulary do we need
008		una festa eh?
009	S:	a party
010	L:	a party and what adjective do we
011		need

Extract 2

C = Charlotte; S = Simona

001	C:	and er Simona picture E
002		(3.0)
003	S:	she:: (.) gets
004	C:	good yeah she gets to work
005	S:	she gets to work at er five to
006		nine
007	C:	good

In particular, what is noticeable here is the way in which Linda elicits language from her students. She uses a pattern of proffering repeated elicitations moving from a more open question to a very narrow one, where the learner is often required only to supply one word as an answer. Moreover, these initiations are then put together in a sequence in order to build up gradually the vocabulary necessary to describe the picture using the target grammar point. In fact, once the learners have the language they need, the elicitation to obtain the target sentence can be more open:

Extract 3

> L = Linda
> 001   L: right using so let's do a sentence
> 002   with so first what will it be

Linda is giving her learners a lot of support, initially guiding their answers and then gradually removing the support as they are ready to use the language alone.

Charlotte, on the other hand, asks the learners to simply describe the pictures, leaving the answers relatively more open and therefore increasing the possibility of the need for correction and negotiation. Clearly, Charlotte's learners are being left freer to experiment and use the language (albeit in a controlled manner) and in fact their answers tend to be longer and relatively more complex and we have evidence of them experimenting, as the following example shows:

Extract 4

> LL = unidentified learners; N = Nino; M = unidentified male learner;
> C = Charlotte; S = Simona
> 001   C: and er Nino C
> 002   N: er she don't er
> 003   aspetta (mi fai provare)
> 004   *wait (let me try)*
> 005   she don't er er ah she don't (not)
> 006   has er breakfast
> 007   C: ok and erm
> 008   N: or she hasn't breakfast
> 009   ((sound of chalk))
> 010   N: she hasn't
> 011   LL: ((mumblings of 'doesn't'))
> 012   N: she doesn't
> 013   ((sound of chalk))
> 014   M: she doesn't have
> 015   ((sound of chalk))

016	C:	she
017	S:	she doesn't
018	C:	She doesn't yeah
019		((sound of chalk. LLs repeat 'she
020		doesn't' quietly))
021	C:	she doesn't have breakfast

From these examples it can also be seen how the teachers assign turns differently. Linda rarely nominates a particular learner to speak. Again, this contributes to the 'smoother' interaction in Linda's extract as the learner who self-selects is more likely to proffer a correct answer, thereby allowing a simple three-part IRF sequence to take place:

Extract 5

L = Linda; K = Katia

001	L:	who is it
002	K:	somebody knocked the the door
003	L:	somebody yeah knocking at the door
004		but who is it probably
005	K:	er un vicino
006	L:	((*laughs*)) the:: neighbour ok

In fact, not nominating is a conscious policy on Linda's part, as she explains in an interview:

> no I don't nominate...I mean because I'm not really interested in the answer I just want them to participate I mean you know amongst all the 14 people or whatever somebody's going to give you the right answer so you pick up on that it doesn't really matter who it comes from

Charlotte, on the other hand, nearly always nominates a particular learner to answer. The advantage of this is that all the learners are given a chance to participate, but the risk is that the nominated learner has difficulty in answering the elicitation, leading to longer and more complex IRF sequences. For example:

Extract 6

C = Charlotte; F =Federica

001	C:	ok so she gets up late er picture

```
002                    B e::::r Federica
003                    (1.0)
004        F:          ermm ... er
005                    (2.0)
006        C:          sorry cause
007                    you ⌈you didn't (xx)⌉
008        F:              ⌊aspetta    er::: ⌋she make a
009                    erm no she go?
010        C:          yeah ⌈she ⌉
011        F:              ⌊(xx)⌋ she go=
012        C:          =she has ⌈a shower ⌉
013        F:                  ⌊she has= ⌋
014        C:          =she has a shower
015        F:          she has a shower shower
```

The result of this, and of similar episodes such as Extract 4 above with Nino, is the need for extensive negotiation of meaning and repair sequences. This means that the interaction proceeds much less smoothly, but on the other hand, Charlotte is perhaps ensuring a more equal distribution of turns and avoiding domination by certain learners. It is also clear that she has successfully created that positive atmosphere where learners 'aren't embarrassed to make mistakes', which is so central to her beliefs.

The way that Linda keeps control over the interaction can clearly be seen as connected to the focus of her belief system on learning outcomes. By scaffolding the learners' answers and not nominating specific learners to speak, she is ensuring that the interaction moves forward and that the lesson and learning progress smoothly, a central aspect of her belief system. Moreover, controlling the interaction can be seen as part of what she believes her teacher role is, something she is expected to do. On the other hand, Charlotte's focus on relationships and creating a positive affective environment is reflected in the greater space she allows her learners and her relative lack of concern with smooth progress through the lesson. This creates an atmosphere in which her learners can take more risks with their language use.

Teacher follow-up

Like their initiation slots, the patterns that the two teachers use in their Follow-up moves are also significant.

Both teachers are quite sparing in their explicit evaluations, and nei-ther teacher uses explicit negative evaluations. Both favour the tech-nique of accepting the learner's contribution by saying *'yes'* or *'ok'*, whether that contribution is correct or not. Although such F-slots are clearly evaluative (Cullen 2002), the fact that the teachers tend to use *yeah* or *ok*, rather than explicit evaluation, could be more encouraging to learners (Nassaji and Wells 2000), something that both teachers express concern with in their interviews.

Linda's F-slots again tend to follow a pattern and often consist of three elements – she accepts the answer (*ok, yes*), sometimes repeats it and then elaborates on it, thereby giving the learners some more infor-mation about the language. This is consistent with her beliefs, accord-ing to which part of her role is having knowledge about the language and to transmit this knowledge to the learners. The following example is the continuation of Extract 1:

Extract 7

L = Linda; M = Marco

001	L:	a party and what adjective do we
002		need
003		((general mumbling from LLs))
004	M:	loudly
005		((general mumbling from LLs))
006	L:	loud or loudly
007	M:	ah no loud loud
008	L:	loud yes it's an adjective we can
009		say loud or we can say noisy ok

Similar to Linda's tendency to add extra information about the language in her Follow-up slots and to use metalanguage to do this, is her use of metalinguistic feedback in correction. The following is an example:

Extract 8

L = Linda; G = Giovanni; M = Massimo; K = Katia

001	L:	ok so (xxxx) so loud so you
002		have noisy so plus adjective um
003		let's just change the first bit
004		with such
005		(1.0)

006	L:	what does it how can you say it
007	G:	he he
008	M:	it was a
009	K:	it was a
010	M:	such loud party
011	K:	such loud and noisy party
012		(2.0) ((sound of chalk))
013	K:	that er
014	L:	that the neighbour complained
015		right it was such loud party um
016		something's missing what's missing
017	M:	a such loud
018	L:	careful look at the rules is it a
019		such not exactly
020		(2.0)
021	M:	such a loud
022	L:	yes that's right such goes before
023		the erm article goes before a so
024		it was such a loud party um ok

This type of feedback and correction is absent from Charlotte's lesson. Both can be seen as consistent with Linda's belief in the importance of the focusing on the language itself as subject matter and her belief in her role in giving learners knowledge about that language.

Charlotte's F-slots tend to have only two parts. She also accepts the learners' answers (*yeah, ok*) and then gives the complete answer, which is sometimes repetition, but more often it is the completion of a partial answer. In fact, Charlotte rarely insists that the learners produce the answer that she wants, or indeed a complete answer. She will use her F-slot to supply the correct version for the learners, as happens in line 017 in Extract 9:

Example 9

C = Charlotte; M = Mirco

001	C:	ok now with your partner (.) can
002		you (.) repeat Suzy's day ok so
003		let's just do (.) an example
004		Mirco, can you tell us picture A

005	M:	uhu (xxxx) get up er late
006	C:	yeah tell us (with) Suzy Stressed,
007	M:	Suzy's stressed because er she get
008		up late
009	C:	she,
010	M:	er (.) cos'e'
011		*what is it*
012	C:	sss
013	M:	get vuole s
014		*takes s*
015	C:	yeah she gets
016	M:	gets si si yes
017	C:	ok so she gets up late

Again Charlotte is less concerned with eliciting correct answers from the learners than with creating a positive affective atmosphere by getting their participation but not insisting on accuracy. The learners are given the input via her modelling the language for them, without further explanations or use of metalanguage.

One final point to note is the way in which Charlotte will allow the learners to take some control of the interaction. This can be seen in an episode which starts with Extract 10.1 below when the teacher's 'ok' in line 001 would seem to be what Sinclair and Coulthard (1975) call a framing move, that is, a move with the purpose of signalling the closing down of the previous activity and moving on to a new activity. However, two learners respond to it as a genuine question and the teacher takes up their response:

Extract 10.1

X = unidentified male student; M = Mirco; S = Simona; C = Charlotte

001	C:	ok,
002	X:	no
003	M:	no
004	C:	no it's not ok ((laughs)) why why
005		not
006	S:	because is very very stressed
007	C:	yeah also she smokes drinks coffee
008		very bad

This is followed by a series of exchanges, such as the following, that resemble, to some extent, natural conversation:

Extract 10.2

020	N:	coffee Italian is good
021		((laughter))
022	N:	English coffee ugh
023	C:	but English coffee's (xxxx)
024		Italian coffee's very strong
025	N:	strong

This discussion is then followed by a shift back to the IRF pattern as the teacher decides to carry out some language practice and production. This is clearly controlled by the teacher, but is carried out in a personalised context. Therefore her questions are referential and her F-slots are discoursal rather than evaluative (Cullen 2002):

Extract 10.3

026	C:	yeah (.) yeah (.) do you erm do
027		you smoke?
028	M:	no
029	C:	nobody? (.) Federica do you smoke
030		((unintelligible comments))
031	C:	what er what time do you have
032		dinner in the evening

This is by no means an isolated episode in Charlotte's lessons and can clearly be seen as connected to her beliefs. Firstly, her concern with people and relationships explains why she focuses on the learners' own experiences to practice language. Secondly, her acceptance and encouragement of learner initiative can be seen as the result of her desire to satisfy the learners' wants as, by taking the initiative, learners are effectively directing the interaction in such a way that it responds more closely to what they perceive as their needs (see Garton 2002 for a discussion of the positive effects of learner initiative).

This is in contrast to Linda, whose I- and F-slots involve far more teacher talking time and fewer opportunities for the learners to produce language or take the initiative. On the other hand, her learners are being exposed to more language and are being given more information about it than are Charlotte's learners. This is also consistent with Linda's belief in the central role of the teacher in providing learners with input, in keeping control over the learning process and ensuring learners' progress.

Implications

Given the fundamental role that classroom interaction plays in successful language learning, understanding what language teachers do in their classrooms, why, and the effect this has on opportunities for learning is clearly of great importance.

The discussion above shows that Linda's and Charlotte's beliefs about teaching and learning form coherent personal belief systems. Analysis of their classroom interaction also suggests the interaction patterns they set up are consistent with these beliefs and this confirms those of the majority of studies on other aspects of classroom behaviour in TESOL (e.g., Binnie-Smith 1996; Breen, Hird et al. 2001; Johnson 1994; Woods 1991).

What is noticeable, however, is that both the belief systems and, consequently, the interaction patterns of the two teachers are very different, yet both are considered to be highly effective and very successful teachers by both their learners and their colleagues. This means that, as teachers, we will inevitably have different beliefs about teaching and learning and different approaches in the classroom. Concepts such as 'best method' and 'good teaching' should therefore be abandoned in favour of the recognition of diversity in teachers and the idea that 'best teaching' is 'the individually best-next-step for each teacher' (Edge and Richards 1998: 571).

If, as experienced teachers, we are to fully exploit the range of opportunities that classroom interaction offers for learning, then we need to become aware of our beliefs and the effect that these have on this aspect of our classroom practice. We need to find a way of exploring what we do in the classroom in order to better understand it and, if we feel we need to, change it.

In what follows I describe three possible ways in which such investigations may be carried out. The descriptions are necessarily brief, but can be followed up in the references given.

Firstly, we may wish to simply experiment with small changes in everyday classroom practice, what the teachers in Huberman (1992: 131) called 'tinkering'. Such experiments can lead to important insights and be very rewarding. Huberman noted that teachers who

> invested consistently in classroom-level experiments – what they called 'tinkering' with new materials, different pupil grouping, small changes in grading systems – were the most likely to be satisfied later than their peers who had been heavily involved in school-wide or district-wide projects.
>
> (1992: 131)

Alternatively, a more systematic approach may be preferred. This would be the case with Allwright's idea of Exploratory Practice (EP) (see, e.g., Allwright 1993; for a detailed summary of its origins, process and principles, see Allwright 2003). EP is based on developing '*understandings* of the quality of language classroom life' (Allwright 2003: 114). Key to EP is the fact that such understandings should come about 'by finding classroom time for deliberate work for understandings, not *instead* of other classroom activities but by exploiting normal classroom activities for that purpose' (ibid.: 121). Thus there are two central features of EP:

1. a focus on 'understanding'; and
2. the integration of investigation into existing classroom practice
   (Allwright 2003: 122)

Finally, we may wish to go one step further and undertake more systematic research with the idea of bringing about a change in practice, through, for example, action research projects (Burns 2003; Edge 2001). Action Research can be a logical next step of EP, if EP has shown that there may be a need for change in practice (Allwright 2003: 126).

Put very simply, action research can be seen as a continuing cycle of action, observation, reflection and planning, which leads to more action. That is to say, in their professional context, teachers observe what is going on and find aspects of their teaching that they feel are worthy of further investigation. The teacher undertaking action research then reflects on this issue, finds out more about it and develops a plan of investigative action. On implementing the plan in action, the teacher observes carefully what has changed and evaluates whether things have improved or not, and whether professional understanding has developed as a result.

Action research enables the experienced teacher to investigate, in a principled and systematic way, their own beliefs and practice, taking into account their own professional context.

The three ways of investigating practice outlined above could be seen as a sequence, with one following on from the other, or they could equally be seen as alternative starting points for the investigation of beliefs and practice. What they all have in common is that they allow teachers to develop towards being the best teacher they can be and to make changes in their practice, if these are felt to be appropriate

Ultimately, the practical aim of research on teacher beliefs and classroom practice must be to empower teachers themselves. This comes about by enabling teachers to become more aware of who they are as teachers, what they do and why, thereby allowing them to establish their own professional development agenda. As Charlotte says:

> I don't know you just get more satisfaction out of it the more you sort of understand what you're doing.

Note

1. All names of people and places have been changed. Most features of speech, such as micropauses, hesitations, repetitions and so on have been taken out of the interview extracts in favour of a very simple transcription.

References

Allwright, D. 1993. 'Integrating "research" and "pedagogy": appropriate criteria and practical possibilities'. In J. Edge, and K. Richards (eds), *Teachers Develop Teachers Research*. Oxford: Heinemann, pp. 125–35.

Allwright, D. 2003. 'Exploratory practice: Rethinking practitioner research in language teaching'. *Language Teaching Research*, 7: 113–41.

Binnie Smith, D. 1996. 'Teacher decision making in the adult ESL classroom'. In D. Freeman and J. C. Richards (eds), *Teacher Learning in Language Teaching*. Cambridge: Cambridge University Press, pp. 197–216.

Borg, S. 2003. 'Teacher cognition in language teaching: A review of research on what language teachers think, know, believe and do'. *Language Teaching*, 36: 81–109.

Breen, M. P., Hird, B., Milton, M., Oliver, R., and Thwaite, A. 2001. 'Making sense of language teaching: Teachers' principles and classroom practices'. *Applied Linguistics*, 22(4): 470–501.

Burns, A. 1992. 'Teacher beliefs and their influence on classroom practice'. *Prospect*, 7(3): 56–66.

Cullen, R. 2002. 'Supportive teacher talk: The importance of the F-move'. *ELT Journal*, 56(2): 117–27.

Drew, P. and Heritage, J. 1992. 'Analyzing talk at work: An introduction'. In P. Drew and J. Heritage (eds), *Talk at Work*. Cambridge: Cambridge University Press, pp. 3–65.

Edge, J. and Richards, K. 1998. 'Why best practice is not good enough'. *TESOL Quarterly*, 32(3): 569–76.

Garton, S. 2002. 'Learner initiative and classroom interaction'. *ELT Journal*, 56(1): 47–56.

Huberman, M. 1992. 'Teacher development and instructional mastery'. In A. Hargreaves and M. G. Fullan (eds), *Understanding Teacher Development*. London: Cassells, pp. 122–42.

Johnson, K. E. 1994. 'The emerging beliefs and instructional practices of pre-service English as a Second Language teachers'. *Teaching and Teacher Education*, 10(4): 439–52.

Legukte, M. and Thomas, H. 1991. *Process and Experience in the Language Classroom*. Harlow: Longman.

Nassaji, H. and Wells, G. 2000. 'What's the use of "triadic dialogue": An investigation of teacher-student interaction'. *Applied Linguistics*, 21(3): 376–406.

Richards, K. 2003. *Qualitative Inquiry in TESOL*. Basingstoke: Palgrave Macmillan.

Sinclair, J. and Coulthard, M. 1975. *Towards an Analysis of Discourse*. Oxford: Oxford University Press.

Woods, D. 1991. 'Teachers' interpretations of second language teaching curricula'. *RELC Journal*, 22(2): 1–18.

5
Teachers Being Observed: Coming to Terms with Classroom Appraisal

Amanda Howard

Introduction

There are many ways in which practising English language teachers can be appraised, and one of these is by means of observation in the classroom. For many in the profession, classroom appraisal is a very important part of their employment, as it determines career progression. However, relevant literature is limited: mainstream education texts available on the subject tend to focus on preparation for the experience and what to do as an observer, rather than what actually happens in the classroom. In the field of TESOL, the focus of observation literature is usually on strategies used for research rather than for appraisal purposes.

Most practising language teachers are already familiar with observation as a teacher education tool, but this is frequently carried out in a supportive environment as a means of professional development. However, appraisal for experienced teachers seems to have different connotations and in this chapter the analysis of transcripts of classroom and interview data provides some insight into what happens when they are observed. By studying what teachers actually do during lessons and observations, this chapter aims to provide those who are interested or involved in this form of teacher appraisal with further information about the process.

At this point it is important to emphasise that not all teachers experience classroom observation: some engage in peer observations, some are evaluated by their students, and some are asked to compile Professional Development files for the purpose of assessing their own teaching. Other teachers do not undergo any form of appraisal at all. However, for many members of the TESOL profession, observation is a part – perhaps not a very welcome part – of their professional life, and if this does not go well there may be major repercussions in terms of their career path, or even basic employment security. The impact of observation feedback in teacher education has been discussed in Chapters 1 and 2, and it is now relevant to investigate the discourse of the observations themselves within the context of a teacher's professional life.

My own interest in observed lessons arose when colleagues approaching an appraisal observation would ask me for help in ensuring that they produced a lesson that their supervisor (the appraisal observer) would approve of. As this supervisor was also involved in initial teacher training, it was assumed that the teaching model presented there would generate the most approval. From this assumption is derived the term *model lesson* that will be used in the rest of this chapter to describe one that has been specially prepared for the purpose of being observed. A normal lesson, that is, one where only the teacher and students are present in the classroom, will be described as being *pedagogic*.

Appraisal is something that goes on all the time in educational settings: teachers constantly appraise the work and demeanour of their students/pupils/colleagues, and they are also appraised by themselves and those that they come into contact with during their working lives. The term *appraisal* has been chosen in preference to *evaluation*, which is too close to *assessment*, an 'old-fashioned' approach 'whereby a superior passed judgement, in isolation, on the personal worth of a subordinate' (Hancock and Settle 1990: 3). This 'judgement' is usually carried out when the superior observes the teacher. Therefore observation (also referred to as interaction analysis by some authors, e.g. Flanders 1979) could be defined as being 'the purposeful examination of teaching and/or learning events through the systematic processes of data collection and analysis' (Bailey 2001: 114). However, in this chapter we shall be focusing on the classroom impact of these systematic processes of data collection, by means of the analysis of transcripts of interaction, recorded during both pedagogic and model lessons. Where the term 'mainstream education' occurs in the text, it is used to refer to sources that relate to primary, secondary or tertiary education, but are not directly related to TESOL. The terms *supervisor* and *observer* will be used interchangeably, as they describe the same person in this particular context.

Teacher observation

Bailey (2001) suggests that there are four main reasons for observation to be carried out in a classroom:

- Pre-service observations by teacher educators
- Observation by novices or colleagues
- Observation carried out by supervisor or head to judge the extent to which performance adheres to teacher expectations
- Collection of research data

In this chapter the focus will be on the third option, as we are looking at what happens in the classroom when practising teachers are appraised. However, many of the findings can also be related to the other categories. As argued by Wragg and colleagues (1996) and other authors in this field, because what teachers do most is teach, in order to assess a teacher it is necessary to assess their teaching. This is generally carried out by means of an appraisal observation, when one or more additional individuals enter the classroom, and watch and record various aspects of the lesson as it progresses. Salient features of the observed lesson are usually recorded and are discussed with the teacher at a subsequent feedback session. As mentioned earlier, many of the literature sources on teacher observation come from mainstream education (Allwright 1988; Croll 1986; Montgomery 2002; Stubbs and Delamont 1976; West 1992; Wragg 1994), although, according to Allwright (1988), there is also a history of observation in ELT. Flanders was involved in the early stages of mainstream interest in the 1960s when he carried out an investigation into teaching styles, and benchmarks were established for observation (interaction analysis) at this stage. However, when he wrote his analysis of teaching behaviour in 1979, Flanders was already anticipating that these standards would become obsolete at some point in the future.

From the books written on the subject, originating in several English-speaking countries, it could be assumed that teacher observation for appraisal purposes has been a common occurrence. However, contrary to popular belief, it seems that observation in mainstream education is not actually carried out quite as often as might have been thought. Writing in 1996, Wragg and colleagues found that only about 28 per cent of teachers were observed at that time, often for only 20 minutes, and it was up to them to decide on the focus of the observation. An Ofsted government report published in the United Kingdom in 1997 identified teacher appraisal as a problem and argued for the use of observation in teacher assessment (Montgomery 1999). Again referring to mainstream education, Marriot suggests that 'the specific observation of the teachers' work has been one of the less carefully analysed aspects of education' (2001: vii). Her focus is on the fact that teachers are often unaware of their strengths and weaknesses, 'because they have not been systematically and constructively debriefed' (ibid.: 9), and that teacher observation must be carried out for the benefit of the school as a whole. This is a somewhat holistic perception of the process, but her emphasis on feedback is interesting. Feedback is indeed a very important area in terms of professional development and future appraisal for the teachers involved, and research has been carried out into the way that observed teachers are debriefed in TESOL educational settings, as discussed by both Kurtoglu Hooton and Copland in this volume. However, space limitations mean that the area of observation feedback for practising teachers will not be addressed in this chapter.

Referring to mainstream education, Wragg (1987) suggests that observers are typically looking for three things, which coincide with the focus of the current study:

- The behaviour and experiences of pupils
- The behaviour of the teacher
- Teaching outcomes

He describes some research into classroom observation that took place in the 1970s and is in line with my own findings. Curious to know what effect an observer might have on classroom behaviour, one investigator (Samph 1976, cited in Wragg 1987) installed microphones in classrooms and then sent in observers, either announced or unexpected, several weeks later. He discovered that, when observed, teachers tended to ask a large number of questions, use more praise and make greater play with pupils' answers than when no one else was present. One of the conclusions from this research was that teachers can react to visitors by second guessing what prejudices and beliefs the observer has and then trying to provide a model which will elicit approval, or obversely, by being independent and doing whatever they would have done had nobody appeared. Some teachers will simply behave in a more pupil-centred way than usual, as in the Samph study, and others may feel more obliged to 'perform' so that their skills are on public display.

Viewed positively, such observation does have an impact beyond the events that occur in the classroom, as Allwright notes:

> Classroom observations need to be valid not only as accurate records of classroom observation, but also, and in a sense more importantly, as records that properly focus on aspects of classroom behaviour that we know to be causally related to learner achievement.
>
> (1988: 44)

Learner achievement is, after all, the main focus of the classroom appraisal process, and this is certainly something that should be borne in mind throughout any language lesson. However, other views of classroom observation are more negative. Bennet's perceptions of the use of observation as an appraisal technique reflect a rather less comfortable side:

> such a system turns every appraiser into a judge and jury. The appraiser decides what constitutes 'good' teaching, based presumably on their own years of experience; the appraiser selects the criteria upon which the final judgement is made; the appraiser represents the 'management' in providing a control mechanism to keep the profession in line. The appraisee's role is very much limited to teaching and awaiting judgement.
>
> (1992: 41)

What are classroom observers looking for?

The discussion above provides a small sample of some of the available texts on teacher observation, and seems to suggest that this is an area where opinions can differ considerably. As discussed earlier, observers can be either looking for a predetermined set of performance criteria, or using a checklist to classify teacher behaviour. However, there is a danger that such judgements may be very subjective, as teaching expertise is a difficult thing to measure. So how do we define what is effective in teaching? One argument is that we will recognise it when we see it, but because this is not measurable it makes comparability or transferability problematic. This might explain why there are so many examples of observation sheets available (Allwright 1988; Marriot 2001): to enable observers to decide which are most relevant to their particular situation.

The actual process of teaching a class is highly complex, and in order to be able to make observations about the way in which an individual teacher behaves in the classroom, we need to assume that he or she is guided by certain knowledge and beliefs. Teaching provides visible evidence of this (see Garton, this volume). Recent research has been surveyed by Borg, who comes to the conclusion that teacher cognition is 'the unobservable cognitive dimension of teaching – what teachers know, believe and think' (2003: 81). However, he is describing the unobservable dimension, and in this chapter we are more interested in the observable aspects. It is surely teacher cognition that informs the actions of teachers as they take their classes – for example, teachers are generally uncomfortable about teaching subjects that they are unfamiliar with, as well as being reluctant to employ teaching methodology that they believe will be unsuccessful in their particular context. So it would seem reasonable to assume that their usual, everyday pedagogic lessons are informed by their knowledge, belief and thoughts about teaching. However, does the same apply when teachers teach a model lesson? What is the observable result there of their knowledge and beliefs? It could be argued that in order for an observer to consider that a teacher has shown her or himself to be effective during a particular lesson, both teacher and observer should have similar cognitive values. Therefore it could also be argued that the most successful model lesson would be one where the teacher's actions coincide to a greater extent with those of the supervisor. And if this is indeed the case, the basis for pedagogic decisions in model lessons is very different from that which applies to pedagogic lessons.

During the IATEFL conference in Brighton, England, in 2003, I held a workshop entitled 'Model lessons: to teach or not to teach?' which was attended by almost equal numbers of English language teachers, and those who were responsible for carrying out observations (Howard 2004). The room was separated into these two groups, and each was asked to provide a wish list as to what should be involved in a model lesson. Table 5.1 shows the features that were identified.

As can be seen from Table 5.1, although there are some similarities between the criteria, teachers and observers did seem to be looking for rather different things. However, during a model lesson the onus is on the teacher to provide the type of lesson that the supervisor is looking for, because the latter is carrying out the appraisal and has the ultimate decision as to the teacher's suitability for future employment. Therefore it is important that the observer is able to maintain objectivity, although they, like teachers, are only human, and any number of additional personal and organisational factors may come into play during the appraisal

Table 5.1 Observation wish lists. From Howard 2004

Supervisors' Criteria	Teachers' Criteria
Teacher involved in choice of observation focus	Part of professional development
Material exploitation	Real situation and context
Teacher demonstrates decision making abilities	Clear purpose
Classroom management	Minimal disruption to the class
Rapport	Observer well prepared
Student involvement	Observer involved in lesson
Students learn something	Transparency
Teacher awareness of positive and negative features of the lesson	Teacher has time for post-lesson reflection and the discussion with the observer takes place as soon as possible afterwards
Teacher's ability to grade their language	

process. The research described in this chapter throws light on how these factors impact on model lessons.

Data collection methodology

The research data used here were collected from English language class-rooms in a tertiary college system, and the teachers involved were from North America, the United Kingdom and Australasia. They had differ-ing approaches to classroom observations, which were generally carried out once or twice annually, with the ultimate aim of providing manage-ment with information relating to contract renewal. This meant that appraisal in these circumstances was an activity to be taken seriously, and the teachers agreed to participate in this study as part of their own professional development. This research does not judge the classroom interaction, but it does aim to analyse how the teachers represent their knowledge and beliefs in order to provide information about the obser-vation process.

It was important to establish with some precision what happens in a classroom during a lesson observation, and in order to do this a method was needed to accurately record the interaction between students and teacher. The recent genesis of the MP3 player was timely, in that it allowed the teachers to carry or place the recording device somewhere convenient, and the size meant that it was relatively unobtrusive. Much previous classroom research has involved the obvious presence of a large cassette recorder which could affect both student and teacher behaviour because of its higher profile.

Each teacher taking part in the research was asked to provide a recording of a pedagogic lesson, and another made during a model lesson. I asked that this observation would be one that was being carried out for appraisal purposes, but how and when they did this was left up to them. Some participants did express concerns when asked to provide a sample of a pedagogic lesson, as they said that there would not be very much teaching as they had a lot of administration to get through, but they were told that this would not be a problem – they just needed to provide a typical example of a normal lesson. These comments and the teachers' concerns regarding the amount of administration needed during a pedagogic lesson provided the first clue as to possible differences between this and a model lesson.

The lesson recordings were transcribed, allowing comparisons to be made between each teacher's model and pedagogic lesson, and also between the lessons of the various participants. Because the aim was to discover patterns in the presence or absence of particular features in the lessons, the transcription has not been carried out in the great detail that is required in the Conversation Analysis tradition. However, some of the main features of speech have been included in order to clarify the context. In order to further explore the teachers' thoughts about model lessons, interviews and focus groups were set up after the recordings had been completed to investigate why the informants behaved as they did in the classroom. These recordings were also transcribed. In addition, several supervisors were interviewed about their approach to model lessons, in order to gain some insight as to what it was that they wanted to be able to observe. Again, these interviews were transcribed, and a limited sample of this data is discussed below.

The model lesson: the observer's point of view

At this stage is would be useful to know what the supervisor looks for in a model lesson. The majority of appraisal observers have been provided with some form of institutional checklist, but what is actually recorded would seem to depend on the way in which they personally view the observation. Table 5.2 presents some key findings from interviews with supervisors that illustrate this clearly.

In order to delve more deeply into some of the comments summarised in Table 5.2, the transcript below is taken from an interview with Andrew and demonstrates why an observed lesson might be seen as a

Table 5.2 Key findings from interviews

Supervisors talking about appraisal observations	
Andrew:	Uses self-evaluation, lesson observation and student evaluation, and forewarning of evaluation is required. He wants to see 'bells and whistles'; a display lesson. He says that teachers are often cautious in delivery, but need to make a show of what they can do. The majority of model lessons (70 – 80%) are mediocre, but supervisors like to be surprised and challenged.
Ereberto:	Looks for classroom presence, the ability to get the students to work, and flexibility. The main question is 'are they learning?' He does not believe in '1 way' of teaching – good teachers adapt the lesson as required.
Michael:	Model lessons cause stress in the staffroom – being observed is very high on the 'panic list' for teachers and applies a lot of pressure.

challenge by some teachers. In this interview he was asked about the way in which observations were organised in his institution.

Extract 1

A = Supervisor; I = Interviewer

001	A:	I know the teachers are er um
002		evaluated without – or lesson
003		visits are undertaken without
004		notice being given. So a a
005		walk in er observation could
006		occur with a teacher who doesn't
007		know that they are about to be
008		observed. And er I struggle with
009		the professional aspects of that
010		coming from a hum hum UK
011		educational background myself um
012		I also believe that um that if
013		the teacher is delivering a
014		lesson, then, er, the teacher
015		should be given the opportunity
016		or chance to show their skills
017		and to show what they're they're
018		made of, if you like, er, in the

019		classroom. And, er, as an old
020		colleague of mine used to say,
021		it's time to get the best china
022		out.
023	I:	((laughs))
024	A:	And, er I don't believe teachers
025		should just perfunctorily go
026		ahead with a normal class if he
027		or she is being observed. I
028		think that is rather um doing
029		themselves a disservice and I
030		think it's not showing a great
031		deal of respect to the evaluator.

(April 2006, Supervisor data)

From Extract 1 it can be seen that this particular supervisor has very clear ideas as to what he expects to see in a model lesson. The emphasis here is on advance lesson planning and structure, and achievement of aims, and this is perhaps why he seems to favour observations where the teacher has been given previous notice of the appraiser's visit. Andrew sees the model lesson as an opportunity for the teacher involved to display a variety of their classroom skills (lines 21–22) and argues that they should provide some sort of performance. If the teacher teaches a pedagogic lesson, he suggests, they are not only letting themselves down, but also failing to show sufficient respect for the appraiser (lines 24–31). From the summary comments in Table 5.2, it appears that Ereberto's attitude towards model lessons might be somewhat more flexible. However, he also seems to be looking for 'classroom presence', which could be interpreted in a number of ways. He claims that good teachers adapt the lesson as required, which would seem to suggest that he might not be particularly pleased to see a model lesson derived directly from the course book. Therefore from a teacher's perspective there seem to be several arguments in favour of knowing something about your observer and her or his cognitive beliefs, in order to be able to provide visual evidence that they coincide with your own teaching practices.

Examples from the lesson data

We now need to discover how the observer's requirements are reflected in the actual teaching that takes place. As the majority of teachers observed for appraisal purposes have been able to continue their careers in this institution, it is reasonable to assume that they are providing the observers with the type of teaching that they want to see. But what do the teachers involved think this is, and how might they alter their teaching style in order to make their model lessons more 'observer friendly'?

To better understand the lesson dynamics it is helpful to look first at transcripts of a teacher's pedagogic lesson style, and then at those of a model lesson, in order to establish commonalities and differences between the two. Space limitations mean that it is not possible to provide too many examples, so the extracts below will focus on the way in which lesson structure, correction, confirmatory feedback and questions are used in the classroom by one particular teacher, Teresa. In both lessons the students are working on the topic of weather: but when they respond to a question or prompt it is not always possible to clearly identify individual answers.

Extract 2

T = Teresa; S = Student/s

001	T:	H. So besides the number 1, write
002		the letter H (3.0) OK? Don't draw
003		lines because lines are too
004		difficult to read later, so write
005		the letter H beside number 1 (1.0)
006		and now please try to do number 2
007		to number 15 (1.0) I'd like you to
008		try and do this on your own
009		please. (2m48) Can you write your
010		name, please? Henry (2.0) please?
011		I see lots of papers with <u>no</u>
012		<u>names!</u> (5.0) Put your name down
013		there please.. Put the middle
014		down there Henry. It will be
015		easier. (10.0) It's easier if once
016		you've <u>used a letter</u>, you have
017		finished with it, you put an X
018		through it (1.0) so you can see
019		more easily which ones you
020		haven't used yet. (29.0) Adrian,

021		mm mmhh (2.0) – I want you to try
022		it on your own (2m49) Are you
023		finished?
024	S:	No
025	T:	Then what are you doing? (40.0)
026	S:	(xxxxxxx)
027	T:	Doesn't look like it! (9.0)
028		Arnold, please tune in when you've
029		finished (12.0) Have you finished?
030	S:	Was er not hard
031	T:	Was not difficult
032	S:	It was un difficult
033	T:	OK, when you've finished with H,
034		count out. When you've finished
035		with A, ⌈count out
036	S:	⌊and B

(April 2006, Pedagogic lesson)

In this first part of this pedagogic lesson extract, Teresa's focus seems to be on getting the students to follow instructions relating to the way in which a written exercise is completed (lines 1–9). These seem to be relatively clear, but she does repeat herself, and stresses the ways in which the exercise can be made easier (lines 4, 15 and 19). Additionally, her repeated use of the word 'please' (lines 6, 9, 10 and 13) implies a certain sense of urgency. She has noticed that several of the students have omitted to write their names on the papers, and even though politeness is maintained, she strongly emphasises this fact (lines 11–12). Her interaction with Henry suggests that if she does find that something has not been completed correctly, then she is more than prepared to follow it up. During the long pauses (lines 9, 20, 22 and 25), Teresa is moving around the classroom monitoring her students, and checking individual papers, whilst also maintaining somewhat dialogic communication with them. However, her focus is very much on individual students, and she is quick to notice if there is a problem, such as in line 25, when Adrian's attention has obviously wandered and he is no longer on task. Her response to his answer is somewhat critical (line 27), and at this stage her interaction with the students seems to indicate that Teresa is eager to continue with the next part of the lesson. Her use of questions (lines 2, 10, 23 and 29) focuses on student performance rather than language use. She corrects Arnold in line 31, but does not respond to his creative use of language in line 33, where 'un difficult' is used as a circumlocution for hard. Praise is not apparent during this classroom interaction.

In Extract 3 from the same lesson some indication is provided as to how Teresa corrects her students:

Extract 3

001	T:	Mm read, read these words (13)
002	S:	(xxxxxx)
003	T:	yes
004	S:	OK?
005	T:	OK – it's not just hot. There's
006		something that makes the de ⌈ sert=
007	S:	⌊ what?
008	T:	=special
009	S:	Cool?
010	T:	No, cool, cool is not hot
011	S:	Sun is hot
012	T:	No, (2) sun is a thing, not the
013		weather. What have you got here?
014	S:	(xxxxxx)
015	T:	Ther-mom-e-ter
016	S:	Thermometer you meant dry.
017		What does this say?
018	T:	Dry means no water
019	S:	No water (2) OK, no water

(April 2006, Pedagogic lesson)

Teresa instructs the students to read a text, and in line 2 they provide a response, unfortunately indecipherable, with which she agrees, but they then ask if this is OK in order to obtain her confirmation (line 5). However, in cases where she does not agree with the answer that the students have provided, she just says 'no' (lines 10 and 12), and then provides the correct response. However, there is evidence here that the teacher is prepared to explore the students' contributions with them in an interactive way, allowing students to take the initiative.

We can contrast the above examples from a pedagogic lesson with an extract from Teresa's model lesson. In Extract 4 she is again giving instructions, but she is also using elicitation and focusing on the appropriate use of weather words. One of the most striking differences between model and pedagogic lessons is that the overall noise level in the classroom is much lower in model lessons. In Extract 4 the student contributions are much clearer and relatively unaffected by background noise.

Extract 4

T = Teresa; S = Student/s

001	T:	I'm not going to give you a
002		number – I want you to see which
003		ones are weather words (8.0)
004	S:	Do I have to do this?
005	T:	What else have you got? No, just
006		do the front page (8.0) Just the
007		front page (2.0) Just the weather
008		words here (4.0) OK, let's have a
009		look in the top left hand corner –
010		there's a small box there at
011		the top – what are the weather
012		words in that small box there at
013		the top on the left?
014	S:	Humidity
015	T:	Humidity
016	S:	Sunrise
017	T:	Sunrise and
018	S:	Sunset
019	T:	Sunset. Are there any other
020		weather words (.) in that box?
021	S:	No (.) C or F
022	T:	Yeah – what does the C stand for?
023	S:	Celsius
024	T:	Yeah. The C is for Celsius (4.0)
025		and the F is for..
026	S:	Fahrenheit
027	T:	Fahrenheit (3.0) What is the
028		little circle in front of the C
029		and the F?
030	S:	Degrees, degrees
031	T:	Degrees. So is that a weather
032		word also?
033	S:	Yes, it is

(April 2006, Model lesson)

One of the first things that can be seen in this excerpt is the evenness of the turns; the interaction seems to be carefully structured into an IRF pattern (Initiation-Response-Follow-up; e.g. teacher question–student answer–teacher evaluation) which was not as obvious in the pedagogic extracts. The students do not seem to be taking risks or making creative use of the language, but are focusing very carefully on the words that Teresa is using. The interaction is less exploratory than in the pedagogic lesson transcripts and there is no evidence of student initiative. The goal of the lesson (familiarisation with the use of weather words in written texts) is also kept very clearly in focus. This would seem to comply with Andrew's earlier comments about the clear provision of aims in an observed lesson.

Student names are not used in this extract, which suggests that the teaching style is somewhat less personal. Teresa is giving instructions (lines 1–3 and 5–13) whilst also using questions to check understanding. She again uses repetition to emphasise points as she did in the pedagogic lesson ('front page' in lines 6–8 and the 'small box' in lines 10–12), but this technique is also used for confirmatory feedback, when she repeats the correct responses, such as humidity (line 15) and sunset (line 19). Teresa still does not praise her students, as this does not seem to be her teaching style, but she does make it clear to them when they have used the correct language by giving a positive response (lines 22, 24). She also listens carefully to what the students are actually saying, and there seem to be few opportunities for them to make mistakes, as her prompts are clear.

Summary

On the evidence of the extracts in the previous sections, it does seem that Teresa may have adjusted her teaching style in order to accommodate the observer present in her classroom during Extract 4. Of course, these are only brief transcripts from two lessons, but my research suggests that the model lessons studied often conform to a more structured method of teaching, regardless of a teacher's nationality or the training that she or he may have received. During a focus group discussion with other teachers after the classroom recordings had been made, Teresa made the following comments:

Extract 5

I = Interviewer; O = Ophelia; T = Teresa

001	I:	But formal observations?
002	O:	Formal observations when people
003		are coming in.

004	I:	But it never crosses your mind
005		that people might be observing
006		you in order to praise you for
007		one of your numerous=
008	T:	=Well you know that they're never
009		really going to praise you at the
010		end of it 'cos that's what
011		they're trained to do
012		((all laugh))
013	O:	That's certainly true but I
014		find it really superficial that
015		that they don't know me as a
016		teacher. They're just coming in
017		to look at my (.) my one track a
018		year
019	T:	Well I think I think that the
020		students are not the same either
021		(.) they find out that your
022		Supervisor is coming to see you
023		and you tell them ahead of time
024		that the Supervisor is coming to
025		see you=
026	I:	=Would you always tell them that?
027	T:	I would! And so they are (.) they
028		are different too! So it's not
029		just you that's different it's
030		the students who are different..
031		And sometimes that's a positive
032		thing but sometimes it's a
033		negative thing 'cos sometimes they
034		get so afraid of saying something
035		in front of the Supervisor that
036		your interactive activities go
037		away out the window

(April 2006, Focus group)

From the extract above, it certainly seems as though both Teresa and Ophelia see observations as an opportunity for the supervisor to critique aspects of their lesson (lines 8–11) and that they feel that their observers do not really know them as teachers. The evidence that model lessons might differ somewhat from the pedagogic is provided when we find out that Teresa always tells her class that an observation is about to take place (19–27), which apparently causes changes in their behaviour. It is interesting to note her comments about the students being afraid to speak in front of the supervisor, and as this apparently impacts on the planned interactive activities, it does seem that this is an area that would benefit from further investigation.

Conclusion

From the small research sample presented above, it seems likely that the model lesson taught in front of an observer for appraisal purposes differs from an everyday pedagogic lesson, and that there are some significant differences between the two. From the observers' point of view, model lessons need to be carefully planned, with stated aims and clear stages, material needs to be exploited, and students should be seen to be taking part. From the teachers' point of view, the observer is going to be critiquing rather than praising, so their aim is to avoid including anything that might incur criticism. Therefore, from the focus group evidence as well as the lesson transcripts, we can conclude that the following are viewed as being suitable actions by the teacher when contemplating teaching a model lesson:

- Prime the students before the observation takes place.
- Provide a clear plan and give it to the supervisor/observer in advance of the lesson. A clear structure is useful.
- Cut down on administrative tasks during the lesson, or remove them altogether.
- Provide clear instructions and repeat if necessary.
- Ensure that turns are clear and concise; the IRF pattern is effective.
- Ensure that there is clear signposting, so that students are aware when they are providing the correct responses.
- Have a clear strategy for error correction, and listen carefully to what your students are saying.

Finally, no strategy is necessarily failsafe, but on the evidence above some understanding by the teacher of the thoughts, knowledge and beliefs of the appraiser would seem to be particularly useful. This might seem to go against the grain for some teachers – after all, most of us believe that we teach in the way we do because that is our chosen methodology and it works most effectively for us. Taking the beliefs and methodological practices of another individual into account is not always easy, but this research suggests that this is indeed what happens during a model lesson.

From the appraiser's point of view, perhaps it might be useful to bear in mind that the model lesson may well be a 'performance' – an occasion when both teacher and students exhibit their best classroom behaviour. And as such it should perhaps be seen for what it is: a demonstration of the skills and techniques that teachers are able to incorporate into a lesson, should classroom circumstances permit.

References

Allwright, D. 1988. *Observation in the Language Classroom*. London: Longman.

Bailey, K. M. 2001. 'Observation'. In R. Carter and D. Nunan (eds), *The Cambridge Guide to Teaching English to Speakers of Other Languages*. Cambridge: Cambridge University Press, pp. 114–19.

Bennet, H. 1992. *Teacher Appraisal; Survival and Beyond*. Harlow: Longman.

Borg, S. 2003. 'Teacher cognition in language teaching: A review of research on what language teachers think, know, believe and do'. In *Language Teaching*, 36(2): 81–110.

Croll, P. 1986. *Systematic Classroom Observation*. Sussex: Falcon Press.

Flanders, N. 1979. *Analyzing Teaching Behavior*. Reading, MA: Addison-Wesley.

Hancock, R. and Settle, D. 1990. *Teacher Appraisal and Self-Evaluation: A Practical Guide*. Oxford: Basil Blackwell.

Howard, A. 2004. 'Model lessons: To teach or not to teach? In A. Pulverness (ed.), *IATEFL 2003: Brighton Conference Selections*. Canterbury: IATEFL, pp. 98–9.

Marriot, G. 2001. *Observing Teachers at Work*. Oxford: Heinemann.

Montgomery, D. 1999. *Positive Teacher Appraisal Through Classroom Observation*. London: David Fulton.

Montgomery, D. 2002. *Helping Teachers Develop Through Classroom Observation* (2nd edn). London: David Fulton.

Samph, T. 1976. 'Observation effects on teaching verbal behaviour'. *Journal of Educational Psychology*, 8(6): 736–41.

Stubbs, S. and Delamont, S. (eds). 1976. *Explorations in Classroom Observation*. London: John Wiley & Sons.

West, N. 1992. *Classroom Observation in the Context of Appraisal*. Harlow: Longman.

Wragg, E. C. 1987. *Teacher Appraisal: A Practical Guide*. Basingstoke: Macmillan.

Wragg, E. C. 1994. *An Introduction to Classroom Observation*. London: Routledge.

Wragg, E. C., Wikeley, F. J., Wragg, C. M. and Haynes, G. S. 1996. *Teacher Appraisal Observed*. London: Routledge.

6
Going Outside the Classroom
Muna Morris-Adams

Fred's story

At the age of 80, Fred enrolled in a beginner German course which I was teaching once a week at an adult education centre. He set out to learn the language with a great deal of enthusiasm, and made reasonable progress. After attending classes for about six months he went on a coach trip to Germany, and in the next class told us how he had ordered – in German – a hotdog from a stall. When he offered to pay for it, he was told by the hot-dog seller that he could have it for free because he had spoken in German. The rest of us shared his delight in what obviously felt like a major achievement in actually using the target language.

Teachers are sometimes lucky enough to share in their learners' sense of achievement when they have used the target language for successful communication outside the classroom. In cases of transactional language use, such as ordering food or asking for directions, success is both satisfying and easily measurable: you get what you ordered or you arrive at the right destination. Where more personal or interactional language use is required, not only is 'success' more difficult to define, but there are no tangible ways for the learner of a foreign language to measure outcomes.

This chapter will examine aspects of informal conversations between native and non-native speakers, discuss what can be learnt from such 'real' discourse, and what implications this might have for teaching. We will start with a look at some of the defining features of both informal and classroom interactions.

Informal conversations

The genre of informal conversation differs from classroom talk, and indeed from talk in other contexts, in many ways. The basic features of informal talk tend to be based on what Sacks called 'the technology of conversation', in other words the 'rules, techniques, procedures, methods, maxims... that can be used to generate the orderly features we find in... conversations' (1984: 413). One of the most important of these conversational mechanisms is the turn-taking system identified by Sacks, Schegloff and Jefferson (1974). However, as Schegloff points out:

> not only is there an underlying ordering to the way in which turns are distributed among participants and constructed by them (even when this ends up sounding disorderly): what is *done* in these turns is orderly as well – 'orderly' in the sense of non-arbitrary and non-random. This is often called 'coherence', and has most commonly been understood to pertain to matters of topicality and topical organisation.
>
> (1999: 409)

Such topical orderliness and overall conversational coherence is what this chapter is concerned with, looking in particular at how this is achieved by non-native speakers.

One significant feature of informal conversation, and one which distinguishes it from classroom discourse is the fact that the talk is not part of an instrumental task, that no final outcome is needed, that the talking is an end in itself. Indeed, Eggins and Slade (1997: 19) define casual conversation as 'talk which is NOT motivated by any clear pragmatic purpose'.

They do, however, also point out that informal talk is the location of a very important social undertaking: 'the apparent triviality of casual conversation disguises the significant interpersonal work it achieves as interactants enact and confirm social identities and relationships' (Eggins and Slade 1997: 16).

As there are no specific goals or tasks to accomplish in such interactions, topic is the driving force that keeps talk flowing. Given that the prime purpose of such conversations is to establish and maintain social relationships, the process of finding suitable topics, and of doing the work of introducing, sustaining and closing them, therefore becomes a vital means to this end. Such work is very much a collaborative project, with topics in conversations being jointly established. How topics are developed and responded to will affect the developing or existing relationship. Kellermann and Palomares observe that '[t]he topics of our everyday talk are important, defined by and defining of our relationship with others' (2004: 331).

Informal conversations, then, are distinguished by the following features: equal rights to participate, a focus on interactional goals, the establishment of coherence in talk, and the management of the topical organisation and content of the talk. How does this type of interaction compare with what learners typically experience in classrooms?

Classroom discourse

The following extract from an authentic classroom interaction will serve to illustrate some of the major differences with informal talk. The activity is one which will be familiar to most teachers: practising the use of the present perfect, and in this case, also providing specific, short answers: Yes I have/No I haven't. It is an activity that is most often conducted in pairs, but here it is teacher-fronted.

Extract 1

T = Teacher; S = Student

001	T:	Have you ever walked 50
002		kilometres? Nick.
003	S:	Eh yes
004	T:	Yes what?
005	S:	Yes I have
006	T:	Yes I have. Have you ever broken
		an arm or a leg? Julia.

(author's data)

We can see that there is an explicit pedagogical purpose here: the practice and production of accurate responses to the questions; a simple yes or no is not considered sufficient, and only a very limited type of response is acceptable. Turn-taking is tightly structured, and controlled by the teacher. The teacher decides *who* speaks *when*, by nominating students in turn. The Initiation–Response–Feedback (IRF) pattern (Sinclair and Coulthard, 1975) predominates throughout the activity, with the teacher initiating and giving evaluative feedback on accuracy, while students take the responding role.

The asymmetry of the interaction is further highlighted by the teacher's control of topics; she decides which topic to introduce when, as well as how much and indeed *what* can be said about them.

We can also see how pedagogical issues take precedence over the interpersonal aspects of the interaction, insofar as meaning and content are largely ignored by the teacher. In a 'real' conversation Nick's 'yes' would almost certainly have been followed up with expressions of interest (or incredulity), or at the very least by questions asking for more information.

Classroom talk is essentially institutional talk, and will inevitably be strongly influenced by the goal-oriented nature of the tasks and activities in which learners are involved. As Walsh points out, 'participants in the EFL classroom are to a large extent restricted in their choice of language by the prevailing features of that context' (2002: 4). This in turn means that there may be limited opportunity for learners to take the initiative (but see Garton 2002) or to make either equal or substantial contributions to talk. It also provides little scope for the practice of the sort of spontaneous, unpredictable and 'messy' interaction which can be found in informal talk.

In spite of the increasingly prominent role that the teaching of communicative skills has acquired in language learning, there is still a surprising neglect of the interpersonal dimension of interactions, both in language classrooms and in current course books. Tomlinson and colleagues (2001) found that course books failed to promote interactional language skills, and my own investigation of pair work activities found that they rarely encourage genuinely interactive talk, leading instead to either very structured dialogues or alternate monologues with communication as a one-way event, that is, talking in turn rather than turn-taking. As for specific conversational skills, these are most often described in terms of 'useful phrases', relating to, for example, appropriate register or how to express functions such as agreeing and disagreeing. Occasional reference can be found to strategies for opening and closing conversations, but there is little information about what happens in between these stages, that is, how to develop, sustain or change topics.

Classroom discourse, then, constitutes a genre of talk which in very fundamental ways differs from the talk which learners experience in the outside world.

Non-native speaker discourse

Learners tend to value opportunities to talk with native speakers, not just to practise or improve their language skills, but also for social purposes. It is, for many at least, a major reason for learning a language. However, communication with native speakers can be a complex undertaking, partly because the rules and conventions of informal talk are so very different from the interaction patterns with which learners are familiar from their classrooms. Many of my own learners of English have said they find such interactions difficult, and Xiao and Petraki (2007) found that 87 per cent of Chinese students studying in Australia experienced communication difficulties. They also stated that 'one of the more notable difficulties was finding a suitable topic to get the conversation going on' (ibid.: 13). Wilkinson similarly reported on the conversational difficulties of students abroad, and concluded that

> these students' heavy reliance on the roles and norms of the instructional environment was limiting at best and often inappropriate in out-of-class conversations.
>
> (Wilkinson 2002: 168)

This reliance manifested itself in a reluctance to initiate topics and an inability to comprehend more subtle topic initiations which did not involve the use of questions.

In the context of conversational competence, Yano, Long and Ross have described the *content* of conversations with non-native speakers as having a '"here-and-now orientation" and to treat a more predictable, narrower range of topics more briefly', while the *interactional structure* is described as 'marked by abrupt topic shifts, more use of questions for initiating moves, more repetition [...], question and answer strings' (1994: 192).

While there has been a tendency for research to concentrate on the difficulties involved in native-speaker–non-native speaker interactions, and to a large extent on the 'deficiencies' of the non-native speaker, a more positive picture of the communicative abilities of non-native speakers is beginning to emerge. Kurhila, for example, found that the non-native speakers in her study could be 'competent interactants, even though they may not be competent speakers of the particular language' (2005: 224). The conversation extracts that follow will show how non-native speakers of English demonstrated conversational competence both in terms of topic management and interactional procedures.

The conversations

The extracts discussed below are taken from a dataset of ten conversations between native English speaker and non-native speaker (NNS) participants.

The NNS participants were all undergraduate students on a one-year exchange programme at a British University, where they attended a one-year English course. They were all in their early twenties; two are male, and eight are female. Both the males are French, and the females German, Turkish-German, French, Slovakian, Belgian, Norwegian and Japanese. Their language levels varied from the lower intermediate range to advanced. Their conversation partners were described by the NNS participants as classmates, flatmates or friends, and were primarily other (English) undergraduates. All the conversations were recorded by the NNSs themselves, without the researcher being present.

The participants had complete freedom of choice with regard to topics and duration. These are, then, as close to being naturally occurring, spontaneous conversations as one can get when a tape recorder is present as an invisible listener.

There was an initial assumption that a range of communicative difficulties would be present in the recordings of these conversations. However, there were very few instances of misunderstandings, and not many communicative difficulties. In four of the conversations the language used by the non-native speakers caused no problems at all; the other six conversations contained only 19 instances of 'linguistic trouble spots' which required some form of negotiation of meaning to establish understanding. In spite of inaccurate and inappropriate use of grammar and vocabulary, it was clear that, in the words of Gardner and Wagner, 'apparent linguistic deficits often are not interactionally significant to either the first- or second-language-speaking participants' (2005: 2).

The non-native speakers were able to make substantial contributions to the talk, and their contributions were both relevant and coherent. There was no evidence of topic avoidance by the non-native speakers, and they did not limit themselves to 'here and now' topics, nor did they change topics abruptly or frequently. In fact, it seemed that the smooth movement from one topic to another was one of the main reasons why the conversations appeared to progress so well, as we shall see below.

Conversational topic

Brown and Yule famously stated that '"topic" could be described as the most frequently used, unexplained, term in the analysis of discourse' (1983: 70). Without going further into the complexities associated with definitions of the concept, in terms of these informal conversations, 'topic' is taken to be a dynamic, multifaceted entity which is at the core of the conversations, and the establishment and maintenance of which is very much a collaborative project. It is defined as stretches of discourse which have an identifiable and sustained focus, and which are bounded by specific moves that lead to a recognisably complete or partial change of focus. Topics also have a role as coherence-organising devices, as reflected in Svennevig's definition of topic as 'a process, that is a set of techniques for establishing boundaries and coherence patterns in discourse' (1999: 164).

There are recognised strategies for establishing topical boundaries, for changing topics or closing them and for introducing new ones, and some of the techniques or strategies employed by the non-native speakers to manage topical movements will be illustrated with examples from the conversations.

Topic initiation

We have seen that students rarely get a chance to nominate topics in the classroom, and yet this is a key skill for both starting and keeping conversations going.

We shall look first at topic initiations, that is, introducing a new topic, as this is the first thing that happens when a conversation gets under way. Button and Casey (1984) provide typical examples of the sorts of general enquiries that may follow the opening components of talk, often along the lines of *What's up?* or *What's new?*, for example. It is, in effect, an oblique way of saying *What shall we talk about?* The topic, in other words, is up for negotiation at the beginning of a conversation.

In the case of these particular conversations, the participants will clearly have engaged in at least some prior discussions about where, when and for how long these conversations should go on. Starting these conversations has consequently had to be a much more conscious decision than would normally be the case, and this in turn may throw some light on participants' awareness of the conventions involved in initiating conversations. We can see evidence of this in the extract below.

Extract 2

J = Joan, native speaker; O = Oda, non-native speaker (Japanese)
001	J:	Hello Oda How are how are you
002		today?
003		((joint laughter))
004	O:	I'm fine thank you. It's a pretty
005		nice day
006	J:	Yeah it's really warm

007	O:	I really wanna go out today but
008		like you I've got a lot of
009		homework as well
010	J:	Yeah=
011	O:	=yeah
012	J:	A whole essay to type two thousand
013		words I don't know how I'm going
012		to do it

(author's data)

The native speaker here employs a standard opening format for the conversation, and the mutual laughter indicates that they are both aware that it is not entirely natural in this case. Oda's next comment functions as a 'topic initiator', and it is a fairly conventional one in terms of conversation starters: namely a comment on the weather. After Joan's agreement and brief comment in line 6, it is up to Oda to build on this contribution in an attempt to establish a more substantial topic. She does this quite effectively by maintaining a link with the previous topic: she has too much homework to enjoy the nice day. The new topic which she proposes, 'homework', is likely to be of mutual interest, and she specifically draws attention to what they have in common by saying 'like you'. Joan shows her acceptance of this as a suitable topic by providing a comment which expands on it, and they continue to talk about essay writing difficulties over the next many turns.

Topic initiations, then, are jointly negotiated to achieve a focus for the talk which is of mutual interest to the participants. This is in stark contrast to most topic initiations in classrooms, which, as we have seen, tend to be introduced by the teacher, and usually take the form of a question rather than a comment.

Topic change

In classroom interactions it is also the teacher who decides when talk about a particular topic must stop, whereas in informal talks, a topic may either gradually run out of steam, or be mutually closed down by the participants. Topic closures are often marked in fairly specific ways. Howe (1991: 1) identified pauses, summary assessments, acknowledgement tokens, repetition and laughter as indicators of impending topic change.

It is quite common to find more than one of these signals of closure, and an example can be seen in Extract 3. Bella is French, and Max is her

English friend who is eating a typical student meal of baked beans with cheese on top.

Extract 3

```
B = Bella; M = Max
001      B:      Whaa:: it's awful
002      M:      No it tastes good though
003      B:      It tastes good ?
004      M:      Yes do you want to try it ?
005      B:      No I don't want to try hehehe (...)
006              So what (..) what did you do last
007              night ?
```
(author's data)

In line 1 Bella expresses her feeling about the look of this particular dish. Max defends the taste, while Bella in return repeats his statement with questioning intonation. This repetition clearly functions to seek confirmation, but may also contain an element of assessment, as in *Are you serious?* – expressing surprise or disbelief that something that looks so awful can taste good.

Max in line 4 provides the confirmation, and by suggesting she tries it, effectively responds to the implied disbelief. Bella laughs and declines, again using a repetition of Max's phrase. This would seem to be an example of what Svennevig calls an 'echo answer', an expanded response 'that repeats elements of the question' (2003: 285). Where such an expanded response is employed, he argues that 'the repetition marks a strengthened affective commitment by the speaker' (ibid.: 286).

So we already have two potential signals that this topic may be coming to a close – a summarising repetition and laughter, and these are followed by a third, namely the pause just before Bella's change of topic.

She starts her topic initiation with the discourse marker *So* as a way of signalling a new stage in the talk. She follows this up with a question, which is a common means of initiating a topic change following closure. A question, of course, requires an answer, and there is therefore good reason to think that it might succeed as a topic initiator, as indeed it does in this case.

In spite of the fact there is no apparent link between the topic of food and that of Max's whereabouts last night, this topic change does not constitute an abrupt topic change. It merely shows an established strategy for topic change where there is a gradual closing down of the current topic, followed by a negotiated introduction of a new one. Topic change, in this instance, is a two-stage process, consisting of topic closure and topic initiation. A distinguishing feature of such topic changes is that there is no connection, either lexically or propositionally with the previous turn.

Complete breaks from previous topics can also occur without the use of any of the strategies mentioned above. Topics can be suddenly interrupted for all sorts of reasons, like phones ringing, children crying, food being served and so on, but also by a need to clarify something that has been said in the talk so far.

This clarification process can have various topical consequences: it may interrupt the topic permanently, and the element causing confusion may itself become the new topic. It may also lead to what is called a side-sequence, where the confusion is cleared up, and the topic is resumed. Alternatively, an entirely new or a related topic may be introduced.

An example of a topic interruption and the consequent clarification process can be seen in Extract 4, which is a continuation of the conversation between Max and Bella. The first line shows Max's response to Bella's question: What did you do last night?

Extract 4

```
M = Max; B = Bella
008      M:      Last night I went to Pounded and
009              got pounded
010      B:      You got pounded?
011      M:      Yeah
012      B:      What what does it mean?
013      M:      It means I got very very drunk
014      B:      Ah ah no you got very very drunk
015      M:      Uh
016      B:      Yeah but it's very strange eh it
017              seems to be the motivation of most
018              of the students when they go out
019              is just to to to be very drunk
```
                                                                    (author's data)

Max uses partial repetition to link his answer to Bella's question, and gives the required information. Bella in line 10 repeats part of the answer with questioning intonation, and she is, in this case, using repetition to seek clarification. This is misunderstood by Max as merely a confirmation check, that is, he is not aware that more than confirmation is required. Bella therefore asks directly for the meaning of the word 'pounded'. The misunderstanding here is, however, soon cleared up with a straightforward explanation by Max. Bella reacts with expressions of understanding *Ah ah*, and *No*, which may express sympathy or surprise, and follows these with another exact repetition of the phrase used by Max. This repetition functions partly to display her understanding, but it may also contain a 'learner' element of repeating new information to help memorise it.

She then picks up on one aspect of Max's explanation, and uses this to initiate a new, more general topic of the drinking habits of English students. She uses the difference in cultural behaviour as an effective resource to move the discussion on. According to Thornbury and Slade, 'one of the main purposes of any language activity is to explore and negotiate *difference*. If there were no difference, there would be no need to communicate at all' (2006: 285–6).

So again we see an example of the non-native speaker using initiative in taking control of the topic of the conversation, and of being able to make new contributions to keep the conversation going.

Topic transition

Where there is clear connection with previous talk, as in the extract above, we get what I prefer to call a *topic transition*. The connections with previous talk can be established either because there is some sort of relevance to the content of what is being talked about, or because there is a lexical link. This technique is by far the most commonly used for effecting topical shifts in conversations, and was aptly described by Sacks:

> A general feature for topical organisation in conversation is movement from topic to topic, not by a topic-close followed by a topic beginning, but by a stepwise move, which involves linking up whatever is being introduced to what has just been talked about, such that, as far as anybody knows, a new topic has not been started, though we're far from wherever we began.
>
> (quoted by Jefferson 1984: 198)

In Extract 5, from the same conversation as Extract 2, we can see an example of such a move; in this case the conversation has been going for a while, and Joan has been talking about her essay, the topic of which is child psychology.

Extract 5

J = Joan; O = Oda

001	J:	so the good and bad points about
002		research methods which is really
003		boring
004	O:	Really boring ? but if you if you
005		become mother like it could be
006		really useful
007	J:	Yeah but not yet ((laughs))
008	O:	((laughs))
009	J:	I'm not planning to become a
010		mother for a long long time
011	O:	Ah I see yeah I'm not sure that
012		I'm going to get married in the
013		future (..) but I haven't got you
014		know I haven't planned any ( )
015	J:	Well (..)I'd like to get married
016		but it's just so hard trying to
017		( ) a decent man

(author's data)

In line 1 Joan makes a summarising comment which contains an assessment. In such cases it is convention for a conversational partner to match or to 'up' this assessment, or at least to comment on it. Oda does this by repeating the assessment exactly, but with questioning intonation, which could indicate either surprise or disagreement. Her follow-up comment helps to clarify her meaning – that it can't possibly be boring because it could be useful in the future. Joan responds with a *Not yet* and laughter, and Oda joins in. In line 9 Joan shows her acceptance of this topic initiation by making an additional comment, and the topic has then moved from talk about essays to talk about getting married.

It is the non-native speaker who takes the initiative to introduce a new topical direction. She uses repetition to effect this, which maintains coherence, while also showing interest in and alignment with her conversational partner, and together they jointly negotiate the new topic.

The next two extracts, both from the same conversation between Laura from Slovakia and her English friend, Claire, show how topics move from talk about clothes to pocket money to train travel to discos, all without any obvious topic-changing signals. They have been talking about the problems associated with mothers buying clothes for them, and Claire then explains that she started buying her own clothes with pocket money.

Extract 6

```
C = Claire; L = Laura
001    C:    pretty much (..) If my Mum's going
002          to buy me clothes ( ) for
003          Christmas but ever since like I
004          started getting pocket money I
005          used to save to buy clothes
006    L:    Uh hu (...) I never really got
007          pocket ⌈ money ⌉ every week I only
008    C:         ⌊ Oh    ⌋
009    L:    got money when I needed it
```
(author's data)

Laura uses lexical repetition to pick up on the issue of pocket money, contrasting her own experience with that of Claire. She therefore maintains conversational coherence, while still succeeding in initiating a new topic. Claire's response is crucial in whether or not this gets accepted. *Oh* generally signals surprise at new or unexpected information, and if we hear something unexpected, we generally want to find out more about it, so it also functions as encouragement to continue. Laura continues to elaborate on the pocket money situation, and includes a related anecdote about how she would sometimes fall asleep on the train on the way to school.

The next extract starts with Laura summarising this experience, to which Claire responds with a sympathetic agreement and a follow-up question in lines 7 and 8. Laura answers the question, and Claire maintains her sympathetic stance and provides an evaluative comment. Like Laura earlier she then provides a personal comparison, using partial repetition of Laura's answer: getting up at quarter to five. She contrasts this with not getting to bed early, and again Laura shows alignment by explicitly drawing attention to this in line 15 with the expression *it's the same with me*; in other words, they share similar experiences which helps to create understanding and to consolidate their friendship.

Extract 7

```
L = Laura; C = Claire
001      L:      I was able to fall asleep for ten
002              ⌈twenty
003      C:      ⌊uh
004      L:      minutes no problem (..) when I
005              was tired (..) and I was tired
006              every morning ((laughs))
007      C:      I bet you were. What time were you
008              getting up then?
009      L:      Yeah the earliest was quarter to
010              five
011      C:      Oh my god (..) That's horrible I'm
012              sometimes awake by quarter to five
013              but that's only 'cause I haven't
014              gone to bed yet
015      L:      ((laughs))it's the same with me I
016              mean since sometimes since eh I've
017              I've been going to the discos and
018              so
```

So in these extracts we have seen several different strategies for moving the topic in a different direction, such as the use of repetition to maintain a coherent link while introducing new information, giving an example of the topic under discussion, telling a related anecdote, and, not least, explicitly or implicitly comparing and contrasting experiences. Interactional concerns are therefore strongly implicated in topic transitions and contribute to building mutual understanding.

On the whole, then, these learners would seem to display competent conversation and topic management skills. They can certainly not be described as incompetent communicators.

Lessons from learner discourse

Much classroom time and effort is expended on promoting learners' communicative competence in order to help them to communicate effectively outside the classroom. With greater awareness of the features of informal talk we are in a better position to help them achieve this aim. As Widdowson said: 'The appropriate English for the classroom is the real English that is appropriately used outside it' (1996: 67).

Analysis of learners' out-of class discourse can reveal which strategies are effective in achieving conversational goals, and which might therefore be worth incorporating into classroom teaching. This goes, for example, for ways of showing interactional awareness by negotiating topics of mutual interest, a skill which is not easily practised in classes where the focus is usually more on the outcome of a task or activity rather than on the conversation partner.

Learner discourse can also provide useful indicators of classroom 'gaps'. One such gap relates to the uses of repetition in negotiating meaning and managing shifts of topics. Tannen (1989) has written extensively about the different functions of repetition and how pervasive repetition is in conversations. She points out that

> Repetition enables a speaker to produce language in a more efficient, less energy-draining way. It facilitates the production of more language, more fluently.
>
> (Tannen 1989: 48)

In Extracts 3 and 4 Bella makes extensive use of repetition, and it enables her to make many and coherent contributions to the talk. However, over-reliance on one particular conversational strategy to perform a range of different functions may sometimes have unforeseen and possibly undesirable consequences, such as misunderstandings.

Repetition also plays a prominent role in classrooms, and the functions which it has there may transfer, inappropriately, to other discourse types. Learners need to be aware of the many and distinctive uses of repetition, and also that it is a very useful resource, not just in terms of fluency, but also to link turns and topics in discourse.

Topic management deserves more classroom attention, because, as Riggenbach (1991: 439) observed, the ability to initiate topic changes is an important aspect of conversational fluency. While questions can be a useful means of lower level learners maintaining participation and involvement in interactions, at higher levels learners also need to know, as Wilkinson (2002) suggests, how to use and recognise non-interrogative topic initiators. Ellis points to the possibility that '[...] when teachers allow learners the chance to topicalize, acquisition is more likely to take place' (1999: 222).

While more research is clearly needed to establish the extent to which classroom interaction patterns may transfer to the outside world, a useful starting point for teachers would be to investigate the discourses which prevail in our own classrooms. Walsh argues that 'developing interactional awareness has to begin with the teachers' own data, analyzed by teachers' (2006: 139). Such investigations could, for instance, throw light on the ways in which features of topic management compare with those found in real-life discourse.

Whether or not it is possible or desirable to try to replicate 'real' conversation in the classroom is the subject of much debate (Richards 2006; Seedhouse 1996), but there would certainly seem to be a case for a greater focus on the teaching of conversation skills as one element of communicative competence. The challenge for teachers then becomes one of considering how best to do this.

One approach, advocated by Dörnyei and Thurrell (1994), would be for learners to practise the micro-skills of conversation, such as turn-taking and conversational openers. Another possibility would be to include a space in lessons where learners can talk freely, without teacher interference, without the constraints of structural or vocabulary practice, and more importantly where learners can choose the topics. One such space could be when classes resume after the weekend, when everybody is likely to have something newsworthy to report, or learners could discuss topics they have previously nominated as interesting. Alternatively, the teacher could set a 'starter topic', and then let learners continue, possibly recording the talks and, with learners, analysing relevant discourse features. In addition, studying examples of *successful* learner interactions with native speakers or with other non-native speakers can provide useful and encouraging models of discourse strategies.

It would also be worth exploiting any opportunities which arise during classroom activities to move out of 'classroom mode' and engage in more natural and more personal interactions; for example, by responding to topics raised spontaneously by learners, letting other learners comment, and using backchannels and assessments to encourage them to continue talking. It has been suggested (Ellis 1999; Kumaravadivelu 1993; Morris-Adams 1997; Ohta 1999) that participation in such exchanges may provide opportunities to enrich learners' language development and discourse skills.

If, as Thornbury and Slade (2006: 240) maintain, 'the success of a conversation is evaluated less on its outcome than on the quality of the conversational process itself', then we need to find ways of ensuring our learners can participate in this process effectively.

References

Brown, G. and Yule, G. 1983. *Discourse Analysis*. Cambridge: Cambridge University Press.

Button, G. and Casey, N. 1984. 'Generating topic: The use of topic initial elicitors'. In J. M. Atkinson and J. Heritage (eds), *Structures of Social Action*. Cambridge: Cambridge University Press, pp. 167–90.

Dörnyei, Z. and Thurrell, S. 1994. 'Teaching conversational skills intensively: Course content and rationale'. *ELT Journal*, 48(1): 40–9.

Eggins, S. and Slade, D. 1997. *Analysing Casual Conversation*. London: Cassell.

Ellis, R. 1999 *Learning a Second Language through Interaction*. Philadelphia: John Benjamins.

Gardner, R. and Wagner, J. 2005. 'Introduction'. In R. Gardner and J. Wagner (eds), *Second Language Conversations*. London: Continuum, pp. 1–17.

Garton, S. 2002. 'Learner initiative in the language classroom'. *ELT Journal*, 56(1): 47–56.

Howe, M. 1991. 'Collaboration on topic change in conversation'. *Kansas Working Papers in Linguistics*, Volume 16: 1–14 University of Kansas.

Jefferson, G. 1984. 'On stepwise transition from talk about a trouble to inappropriately next-positioned matters'. In J. M. Atkinson and J. Heritage (eds) *Structures of Social Action*. Cambridge: Cambridge University Press, pp. 413–29.

Kellermann, K. and Palomares, N. 2004. 'Topical profiling: Emergent, co- occurring, and relationally defining topics in talk'. *Journal of Language and Social Psychology*, 23(3): 308–37.

Kumaravadivelu, B. 1993. 'Maximising learning potential in the communicative classroom'. *ELT Journal*, 47(1): 12–21.

Kurhila, S. 2005. *Second Language Interaction*. Amsterdam: John Benjamins.

Morris-Adams, M. 1997. 'An investigation of instances of spontaneous personal interactions in the classroom'. Unpublished MSc dissertation, Language Studies Unit, Aston University.

Ohta, A. S. 1999. 'Interactional routines and the socialization of interactional style in adult learners of Japanese'. *Journal of Pragmatics*, 31(11): 1493–512.

Richards, K. 2006. '"Being the teacher": Identity and classroom conversation'. *Applied Linguistics*, 27(1): 51–77.

Riggenbach, H. 1991. 'Towards an understanding of fluency: A microanalysis of nonnative speaker conversations'. *Discourse Processes* 14(4): 423–41.

Sacks, H. 1971. 'On doing "being ordinary"'. In J. M. Atkinson and J. Heritage (eds) 1984. *Structures of Social Action*. Cambridge: Cambridge University Press, pp. 413–29.

Sacks, H., Schegloff, E., and Jefferson, G. 1974. 'A simplest systematics for the organisation of turn taking for conversation'. *Language*, 50(4): 696–735.

Schegloff, E. A. 1999. 'Discourse, pragmatics, conversation, analysis'. *Discourse Studies*, 1(4): 405–35.

Seedhouse, P. 1996. 'Classroom interaction: Possibilities and impossibilities'. *ELT Journal*, 59(1): 16–23.

Sinclair, J. and Coulthard, M. 1975. *Towards an Analysis of Discourse*. Oxford: Oxford University Press.

Svennevig, J. 1999. *Getting Acquainted in Conversation*. Amsterdam: John Benjamins.

Svennevig, J. 2003. 'Echo answers in native/non-native interaction'. *Pragmatics*, 13(2): 285–310.

Tannen, D. 1989. *Talking Voices*. Cambridge: Cambridge University Press.

Tomlinson, B., Dat, B., Masuhara, H. and Rubdy, R. 2001. 'EFL courses for adults'. *ELT Journal*, 55(1): 80–101.

Thornbury, S. and Slade, D. 2006. *Conversation: From Description to Pedagogy*. Cambridge: Cambridge University Press.

Walsh, S. 2002. 'Construction or obstruction: Teacher talk and learner involvement in the EFL classroom'. *Language Teaching Research*, 6(1): 3–23.

Walsh, S. 2006. 'Talking the talk of the TESOL classroom'. *ELT Journal*, 60(2): 133–41.

Widdowson, H. G. 1996. 'Comment: Authenticity and autonomy in ELT'. *ELT Journal*, 50(1): 67–8.

Wilkinson, S. 2002. 'The omnipresent classroom during summer study abroad: American students in conversation with their French hosts'. *The Modern Language Journal*, 86(ii): 157–73.

Xiao, H. and Petraki, E. 2007. 'An investigation of Chinese students' difficulties in intercultural communication and its role in ELT'. *Journal of Intercultural Communication*, 13. http://www.immi.se/intercultural/

Yano, Y., Long, M., and Ross, S. 1994. 'The effects of simplified and elaborated texts on foreign language reading comprehension'. *Language Learning*, 44(2): 189–219.

Reflections on Becoming Experienced

Maneerat Tarnpichprasert

Background

My career in the teaching profession has actually commenced since I received my bachelor's degree in Education. I began as a teacher in the secondary school in which I had also been a student teacher during the last year of my undergraduate study. At this school, my duties included teaching English as a subject to students from Grade 7 to Grade 10 and being a homeroom teacher for one classroom. The teaching experience and other responsibilities I had while being a teacher in this school have really provided such a great experience for my teaching profession. I was also happy that I could bring what I learnt in the pre-service teacher training course to use in the real classrooms.

I worked in this school for around two years before I decided to leave in order to further my study for a Master's degree in the area of English language teaching. After graduation, I had an opportunity to move from being a school teacher to becoming a university lecturer. This movement brought several changes for my teaching profession including my teaching styles and techniques. Some details will be discussed later in the next sections. After teaching for around two years, now I am on leave to do a PhD in English language teaching. As soon as I graduate, I will go back to continue my work as a university lecturer in the same institute.

From a secondary school teacher to a university lecturer

Despite the fact that I was quite happy being a secondary school teacher teaching secondary students who are, at their age, very lively, active and very close to their friends and their teachers, one of the main reasons that motivated me to become a university lecturer is because I believe that teaching at the university level provides me much more freedom to design my own courses than teaching in a secondary school. When I was teaching at secondary level, I very much liked to bring activities into my lessons. I believe that in language teaching, the use of activities will be very useful in encouraging students to participate in the lessons. Also, it can help to create situations in which the target language is used in authentic communication. However, while teaching at secondary level, I had to follow the curriculum from the Ministry of Education which contains a large amount of subject content. It was likely that the amount of time for teaching and the amount of content we had to teach was not very well balanced. As a result, it was very difficult for me to add more activities into my lessons.

As a university lecturer, on the other hand, I have more freedom to design my own courses and I can therefore bring a lot of activities into my classrooms. Being able to teach in the way that I prefer; in other words, being able to use the teaching techniques that I believe will be useful for my learners, I find that I enjoy teaching and have the enthusiasm to prepare lessons and teaching materials. Based on this issue, I totally agree with Garton's idea that teachers' beliefs in teaching and learning have a great impact on their teaching. Furthermore, in my case, it may even raise another possible implication; it is plausible that if teachers cannot follow what they believe is good for teaching and learning, this may influence perspectives of their careers and may even cause them to leave their profession.

In the light of my teaching at university level, at the moment, there is an issue that concerns me. Before leaving for a PhD, my teaching in the university included only some undergraduate courses. It is likely that undergraduate students, especially the first-year students, still need a lot of help and support from teachers. It is also noticeable that the learning styles of undergraduate students, particularly the freshmen and the sophomores, are still very similar to those of high school students. The teaching style I utilised with this group of students therefore focused more on the learning atmosphere. I believe that teaching techniques helping to encourage students' motivation to learn are very necessary for teaching at this level.

After receiving my doctoral degree, however, I will be expected to teach postgraduate courses as well. What I believe is that the teaching styles utilised for teaching undergraduate students and postgraduate students may be slightly different. For the postgraduate level, although the teaching styles that promote an affective learning environment are still very important in the teaching learning process, the knowledge and professionalism of the teachers may be more emphasised. In addition, it is likely that the higher the level of education the students are in, the more they will be expected to be autonomous learners. The role of the teacher teaching at postgraduate level, therefore, should become apparent more as a source of information, guidance and support.

In view of these issues, my concern is therefore whether my teaching in the two different levels will be different or not, and whether I will be able to provide effective teaching to both levels of education or not. In her chapter, Garton has raised a good point regarding the relationships between teachers' beliefs and their teaching and she has also made the point clearer by providing some interesting and explicit examples from her study. After reading this chapter, I try to think of what I believe in teaching and can come up with some ideas discussed above. I am now more aware of my own beliefs and I am positive that this awareness will be beneficial to my teaching preparation for both of my student groups. However, I may first need to find out whether what I believe in teaching these two groups of students is correct or not.

Lastly, it is conspicuous that the author does not appear to discuss the situation when a teacher teaching two different groups of students believes that these two groups of students should be taught in different methods; how his/her teaching in the two different situations will be. This matter may also be a very interesting subject for further investigation.

From being observed to observing others

As a university lecturer, there is also another issue that worries me; it is about my additional role as a supervisor who observes the teaching practice of some pre-service teachers. Given that I am teaching in the faculty of Education, my duty, apart from teaching, also includes being an observer evaluating the professional practice of my students. (In the final year of their Bachelor of Education programme, the students have to attend the professional practice course.) Actually, during the last year of my undergraduate study, I myself also used to be a student teacher and was observed and evaluated by my supervisor. From that direct experience, I totally agree with the point that Howard makes. She points out that in classroom observation, successful teaching seems to be the one where the teacher and the observer share the same 'cognitive values'; in other words, the successful lesson for observation appears to be the one where the teacher's teaching corresponds with the observer's thought, knowledge and beliefs about teaching.

When I was being observed by my supervisor, I found it very helpful to try to have as much information as possible concerning which area the observer was going to evaluate from my lesson and what kind of teaching the observer thought was effective. When I was a pre-service teacher, I discussed these issues with my supervisor and she kindly informed me the areas she was going to observe such as the teaching techniques, the use of materials and activities, classroom control, students' participation and so on. I could then infer from the discussion the aspects that my supervisor thought were important in teaching and learning, as well as the types of classroom environment she considered as effective teaching. While teaching, therefore, I kept these aspects in mind and tried to teach in the way that I thought my observers would approve.

Based on this experience, when I myself became an observer, I understood the students' need to get information regarding the areas they will be evaluated on and also, perhaps, some idea about what kind of teaching their supervisor considers as effective teaching. As a supervisor, what I normally do is after the first observation which usually occurs after a few weeks of their teaching practice, I discuss with my students their teaching in that particular lesson and also other aspects of their teaching, including the characteristics of their students as well as the teaching methods they think should be suitable to use with those students. After they tell me what they think, it is my turn to share with them what I think and I can also recommend to them some ideas that I think will be useful for their teaching. Then, in the next observation, I see how they have improved their teaching. Normally, the students will be observed four to five times per semester. Personally, in my observation, I focus on the teacher's capability to make the students learn and also his/her flexibility to adjust the lesson and the teaching methods to suit the students. In addition, one of the most important criteria I use for marking their professional practice is the development of their teaching performance.

Another interesting issue that Howard notes is the possibility that the teaching structure or the teaching style teachers utilised in the lesson being observed, or what she calls 'model lesson', may be different from their 'pedagogic lesson' or the normal lesson. This issue reminds me of the observational strategy that one of my colleagues utilised when he observed his students. This supervisor would go to observe his students' teaching practice without notifying them in advance. The reason he did this, as he told me, was because he liked to see 'the normal lesson' that his student teachers teach, not the 'display lesson' or the one that they prepared to present to him. Although there appear to be some advantages in using this observational strategy, I am still concerned that this method exerts too much pressure on my students. It may, then, be very beneficial if the use of this strategy is further explored.

As for my own teaching, since I began my teaching there has not been any formal observation to evaluate it. Generally, there is merely some feedback from my students and my own self-evaluation to assess my teaching.

The use of the communicative approach in my teaching

As a language teacher, I concur that the learners' competency in using the target language can be most effectively improved through the frequent use of that language, in other words, the more the learners utilise the language, the better language proficiency they will have. The point that Morris-Adams makes regarding the use of the target language outside the classroom is therefore very interesting to consider. In addition, the author also emphasises the different features of informal conversation and classroom discourse; it is likely that in informal conversation outside the classroom, the learners' use of the target language appears more authentic. Thus, it is possible that in order to increase students' opportunity to use the target language in authentic communication, they should be encouraged to use the language outside the classroom as much as possible.

In the light of her study about the interaction between native and non-native speakers, moreover, Morris-Adams claims that the capability of non-native speakers to use the target language in their communication with native speakers is normally better than they estimate. The author also interestingly discusses some positive results for non-native speakers' communication with native speakers as well as some techniques and strategies they utilise for topic management. However, she does not appear to explain whether these issues also happen similarly if non-native speakers communicate with non-native speakers or not. I am interested in this aspect because in my context, it seems difficult for my learners to have the opportunity to use English outside the classroom and even less if they would like to use it with native speakers. This is because in my country, English is utilised only as a foreign language and we use our mother tongue in everyday communication. In their real lives, the students can rarely be exposed to the use of the target language, especially with native speakers. For listening and reading skills, it is possible that the students may be able to listen to English conversation or to read some texts in English from many kinds of international media. Nevertheless, for speaking and writing, it is difficult for them to have a chance to use it in their everyday lives.

In order to help them to find as much opportunity as possible to use the language in authentic communication, I try to encourage my students to use English outside the classroom. For instance, when there are some overseas visitors coming to our university, I will encourage my students to take part in welcoming those guests; some students volunteer to become a guide for the visitors. Although it is unlikely that there will be many visitors who are native English speakers, I believe that the use of English with non-native speakers will also be useful for my students' language improvement. In addition, it is undeniable that at present, English is usually used as an international language for international communication and we tend to use English to communicate with non-native speakers more than with native speakers. I therefore think that it will be beneficial if this aspect concerning the use of the target language for communication among non-native speakers is also considered.

A message to teachers sharing the same career phase

When considering my own teaching experience, I think I am now moving from the stage of being a 'new' teacher to the phase of starting to become an 'experienced' teacher. In the first few years of my professional teaching life, there were several times I felt happy that I chose to become a teacher, and at the same time there were several moments that I asked myself whether I should change to another profession or not. The former feeling usually occurred when I found my lessons successful: my students were happy with my teaching and they could reach the objectives of those lessons. The latter one, on the other hand, normally happened when I felt disappointed with my students' learning achievement and their learning behaviour, or when I could not follow what I believe would be good for my teaching. It is possible that some teachers, particularly those who are new to the profession, may also encounter these feelings.

In addition, beginning teachers may have learnt several theories and would like to see how these theories work in practice, or they may have some ideas or beliefs about teaching that they would like to try out in their classrooms. The first few years of their teaching may therefore be used to find their own 'identity' as teachers. They should also take this opportunity to consider what they believe in teaching, to try out those beliefs and to learn from their trial. In doing so, it should be emphasised that being open-minded and being flexible are the main keys to learning new things. Comments from experienced teachers or observers should also be considered as good support. Last but not least, it is very important to ask yourself whether you are happy with what you are doing or not – and not to be afraid of changing

Part III

New Horizons

Paula

Paula has now been teaching for ten years and has developed a considerable amount of experience. She would describe herself as a reflective teacher as she is very much concerned with what goes on in her classroom. Of course, she frequently chats to her English colleagues, both in the staffroom and sometimes socially too. Compared to some of the other groups of colleagues in her institution, the English group are quite cohesive, get on reasonably well and tend to share both ideas and materials.

Occasionally Paula tries to keep a diary, writing down her thoughts and reflections on her lessons in order to gain deeper insights into both her students' learning and her own teaching. She has joined a local English teachers' group and she sometimes attends seminars and workshops there. She also goes to the annual national TESOL conference whenever she can, mainly to get new ideas for her classroom. Sometimes, when she has time, she will read some articles that look interesting in a professional journal, but of course, she rarely has time.

However, all this is very casual. Increasingly, Paula is starting to feel it is somehow not enough. The thought that she might continue to do the same thing until she retires leaves her feeling dissatisfied and thinking that there must be more. She has reached that point in her career, like so many teachers, where she feels the need to explore new dimensions in her profession, perhaps in a more principled and systematic way.

This stage would seem to be very much a crossroads in a teacher's journey through the career cycle. The first two stages have been along a relatively straight, albeit perhaps bumpy, road, but now that road goes off in potentially many different directions and the teacher seeking New Horizons must decide which way to go. This section therefore presents three very different chapters, each one of which explores an alternative way for the New Horizons teacher to develop professionally. Linking all three chapters, though, is evidence of the sort of initiative and opening up of participation that is so important if good practice is to flourish (Coldron and Smith 1999: 721–2) and of the positive impact of sharing such practice with supportive colleagues in a strong professional community (Day, Stobart et al. 2006).

Perhaps the most obvious way forward is through further formal academic study and this is the background for Phil Quirke's chapter. Quirke explores how his website supported a group of teachers during a distance learning Master's programme, while at the same time developing their knowledge and understanding. By analysing the different types of knowledge that teachers using the site demonstrated in their interactions, Quirke has developed a theoretical model of a teacher knowledge cycle. Interestingly, the final stage of the cycle is that of *knowledge provider* as some teachers using the site take their first tentative steps towards Passing on the Knowledge.

This is also very practical chapter, as Quirke identifies a series of Keys to successful interaction between those developing a site and those using it. He also offers clear advice to those wishing to set up educational support sites.

Rather than pursue academic study, some teachers may decide to seek further professional development within their local context, by sharing experiences with other teachers in some way. Steve Mann's chapter takes up the idea that teachers get together to talk, but looks at it from the perspective of developing ideas through exploratory professional talk. Mann shows how the space created by non-evaluative discourse in a Cooperative Development framework allowed a group of colleagues the space to articulate their ideas and move their professional development forward. In particular, Mann explores the role of metaphor in the discourse of the group, and especially the power of extended metaphor, not only to articulate ideas but also to understand the nature of the talk itself.

Finally, some teachers may decide to take a more drastic step and change the direction of their career by making the move from teaching to management.

Keith Richards' chapter traces the journey of a group of teachers who took this step by starting a new language school. Richards takes us through every stage, from how the team came together, to the setting up and consolidation of the school. The story is told in the words of the teachers themselves, but we also see how the teachers developed their roles within the group and how they developed very distinctive ways of interacting. The chapter identifies five different orientations which Richards offers as the keys underlying the group's effectiveness as a collaborative team.

All the chapters in this section focus on a particular group of teachers, but they all, explicitly or implicitly, also give practical suggestions as to how teachers can take forward their professional development at what we might call a crossroads in their careers. They indicate some of the ways that teachers at the New Horizons stage of their careers can take forward their professional development by engaging in different forms of discourse and, as such, these will hopefully resonate with many.

That is certainly the case with Jerry Talandis Jr., whose reflections close this section. Talandis is a teacher from the United States working in Japan and he explores what it means to him to be a teacher who is actively seeking New Horizons. There are clearly strong resonances in all the chapters with Talandis's own experiences to date, but at the same time, they offer him insights into his future professional development too.

Although the chapters in this section are all quite diverse, Talandis identifies common themes of introspection, articulation, communication and collaboration, which enable him to begin to see a way forward.

References

Coldron, J. and Smith, R. 1999. 'Active location in teachers' construction of their professional identities'. *Journal of Curriculum Studies,* 31(6): 711–26.

Day, C., Stobart, G., Sammons, P. and Kington, A. 2006. 'Variations in the work and lives of teachers: Relative and relational effectiveness'. *Teachers and Teaching,* 12(2): 169–92.

7
Supporting Teacher Development on the Web

Phil Quirke

This chapter explores how the worldwide web can be used to support teachers involved in developing professionally through further study at Masters level. The study upon which the chapter is based was a two-year qualitative research project, which examined how teachers interacted with a website, http://www.philseflsupport.com, set up to support their studies.

The chapter draws upon teacher emails and discussion board contributions to show how the four roles that teachers developed through their interaction with the website and the website developer (knowledge-seeker, knowledge-discusser, knowledge-user and knowledge-provider) form a knowledge cycle. As their knowledge develops, teachers become more aware of other gaps in their knowledge and begin the cycle again. This cycle (see Figure 7.1 below) is driven by the continuous theorising of practical knowledge and practicalising of theoretical knowledge (Tsui 2003).

The chapter concludes with some practical advice drawn from the research experience which can help both teachers and teacher educators make the most effective use of web-based support for teacher education and development.

The website

The website http://www.philseflsupport.com was launched in order to support teachers doing further education, especially when they were studying at a distance and therefore potentially isolated from professional colleagues. The aim was to develop the website based on the input and feedback of the teachers themselves. When I first contacted colleagues around the world about my ideas for this website, I received

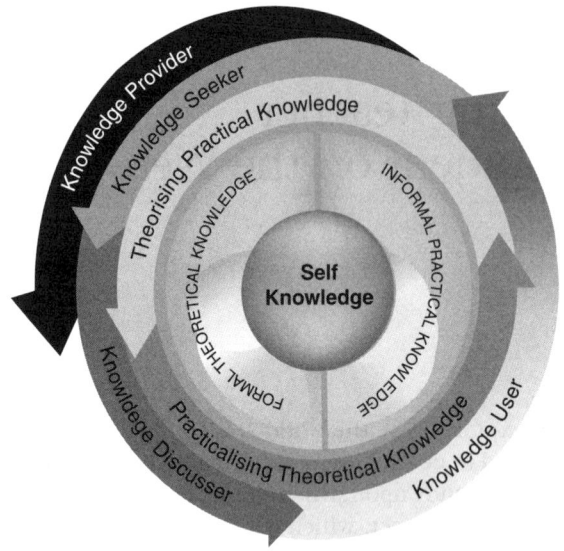

Figure 7.1　Model of teacher knowledge development

109 separate suggestions, which I grouped into cohesive areas. The site was divided into five main areas under a home page, and I then spent three months building the site around this outline ensuring that there was enough content and links to provide good initial support to match the needs identified by those first responses..

The site users' responses and suggestions indicated that the website was fulfilling a need for teachers doing further studies. One year after the launch of the site, a reworked version was uploaded which better reflected the needs of users as evidenced in the first year of feedback. This second version remained predominantly text-based to maintain the speed of the site for all users and especially for those based in remote locations with small band widths. The content was also reorganised to better reflect the feedback from the teachers using the site, so the website structure became, in effect, an informal glimpse into how teachers link schemata of knowledge together.

Subsequently, users began to submit and post their own articles. No changes to the site were made without consulting the users themselves, and I relied on their feedback to continue building the site, so it resembled their needs and work even more closely. This has resulted in a website with over 200 pages, 550 relevant links, 400 registered users and over 13,000 visitors every month. It is a site which therefore seems to be meeting the needs of its target audience of teachers doing further study at Masters level.

Knowledge-seeker

In the model of teacher knowledge presented in Figure 7.1, the starting point is a core of self-knowledge and an initial knowledge of the teaching context based upon our experience. As this experiential knowledge expands we begin to understand the gaps in our knowledge. We seek to deepen this understanding by exploring the theories that underlie our experience through reading, professional workshops and further study; by seeking new knowledge through research and reflection we gain a better understanding of what we are doing in the classroom. This search for deeper understanding as the theorising of practical knowledge (Tsui 2003) defines the role of knowledge-seeker.

The teacher's primary communicative function (Sinclair and Coulthard 1975) in this role is one of knowledge-seeking with the user of the website taking on the role of knowledge-seeker, and the email or discussion board exchange resulting in the user receiving the knowledge sought. The exchanges on the site under this category divided into three turns (Brown and Yule 1983: 230): query (the user requests knowledge), response (to the user request) and thanks (for the knowledge provided). The example below is one of many such exchanges:

> Anyway I was wondering if you have got time, maybe you could give me some advice or some handy hints: (if I knew then what I know now type thing) before I start. You may remember the first module; FND [MSc Foundation Module]. Any feedback/wisdom you might like to share with me would be greatly appreciated as I'm a bit nervous to say the least.
>
> (From User 1.16, 22 August 2002)

> It's a while since I did the FND module and it has changed substantially since then. However, I have had the chance to go through the latest version both for the website and in helping local teachers on the MSc in their approaches. The secret to the FND module is to follow it closely and [*advice follows*]. Please let me know if you need anything from me and I hope this helps.
>
> (Developer to User 1.16, 24 August 2002)

Thanks a lot for replying so quickly & coherently, I'm sure your advice will come in very handy when I start studying.

(From User 1.16, 25 August 2002)

On the Diploma courses I tutor, I refer to the freeze frame in the classroom, meaning we should be able to explain why we are doing what we are doing at any given point in our lessons. This often requires depth of reflection and the seeking of new knowledge to better explain the theories underlying our practice. We are beginning to link our informal, experiential knowledge to formal, professional knowledge using methods such as Wallace's (1991: 39–44) reflection and personal development model.

Teachers using the site commented that it 'promotes reflection and development because it creates an awareness of language within the teacher' (User 1.01) and that 'the activities and language promote reflection' (User 1.04).

Another user commented on how the content was both 'stimulating' and 'challenging' as he described his attempts to match the knowledge required for the worksheet tasks on the site to his existing knowledge structure. User 1.23 provides a good conclusion to this section:

The links are really helpful because they provide quick and easy access to information I need for my studies. The links often lead me to follow up on references or new ideas that I find interesting. Sometimes the information I've found has really enlightened me and precipitated a pursuit of more knowledge in an area.

Knowledge-discusser

The changes to our existing knowledge schema triggered by reflection on newly acquired knowledge are seldom complete until we have involved others. As we shift our schemata, we need confirmation from those professional colleagues around us whom we respect. So we discuss how this new knowledge maps onto our existing cognitive knowledge map and decide whether or not the change we are considering is in fact compatible with our professional context. In this way, the role of professional discussion in the development of our knowledge (Johnson 2000) is an essential element of the model.

The role of knowledge-discusser can be defined as the focus on a specific professional subject over at least four turns and leading to an improved professional understanding for either of the discussers. The number of turns seen in the examples from the research and the constant encouragement for the other party to continue the exchange demonstrate an in-depth search for confirmation and understanding. The use of lexical chaining (McCarthy 1991) and the length and number of turns were characteristic features in the knowledge-discusser role.

One of the most interesting features of the data was the fact that there were so few exchanges which saw knowledge-discusser roles adopted. Most discussions were more social in nature, possibly following standard communication norms of building trust informally before launching into the potentially face threatening forum of professional discourse. It would seem that most professional discussion actually takes place in a face-to-face medium with colleagues at hand and that online forums are mainly used to gather knowledge quickly through direct questioning or lurking, or to provide knowledge to others as a foray into professional discourse and potential publication. The website users also confirmed that they had taken discussions off the site as they set up direct communication with colleagues around the world who shared their research interests.

> I learned from your experience, and by getting me in touch with people who could help me, I was able to accomplish far more than I could have otherwise...sometimes participants can support each other away from the forum. This happened to me, a bit. The forum is like the café where you meet, and then you take it from there.
>
> (User 1.29)

Knowledge-user

One of the key phases in the transformation of our knowledge structure is the practicalising of new knowledge as we confirm our beliefs in practice. This phase usually occurs away from the website although there was sufficient evidence in my data that pointed to a knowledge-user role for its inclusion in the model of teacher knowledge development. The independence of this knowledge use was seen in the lack of an active developer role, and the definition of the knowledge-user role – 'the user demonstrates they (plan to) use the knowledge available from the site' – demonstrates a direct parallel to the practicalising of theoretical knowledge with its focus on action research in the classroom (Wallace 1997) and collaboration (Edge 1992; Woods 1996). In this role, the primary communicative function was one of demonstrating knowledge use with the user reporting on his actions.

The range of different uses reported gave an indication of the versatility of the site and showed how it met the needs of different users. The email and discussion board exchanges gave evidence of how the users researched, tested and examined their knowledge by using the action research course links, quoting the content in research writing and website linking, using the content in their teaching practice, supporting their research studies, and using the site as their affective support mechanism.

This final use of the site as an affective support conduit was particularly noteworthy as it matched one of the primary aims of the site. One user did his research as an isolated expatriate on a small island. His research was based upon cooperative development at a distance and he saw the development of the website as a similar project. He used the site as a motivational tool as he observed it and my research develop. He stated that he used the site as a bibliographical resource and, more importantly for him, was able to draw upon our email exchanges and make a comparison of our uses of highlighting and prioritizing. He felt my developer emails used 'a similar communication strategy' to cooperative development and active listening. He saw this use of the knowledge on the site as providing him with 'a source of affective support' more than anything else since the main outcome for him was a reduction in his sense of isolation.

Another user was based in Abu Dhabi. He used the email exchanges to set up a meeting to discuss the theoretical research I had done and its relevance to his studies. This move from a web-based discussion to a face-to-face forum supports the conclusion in the previous section that much professional discussion will move away from an email or discussion board exchange if the opportunity arises.

Much of this stage of the cycle occurs away from the site since it is practicalising what we have taken away, reflected upon and discussed, so most teacher knowledge-use will take place in the institution or classroom as research, lessons, materials and courses are prepared. This experimental stage of knowledge-use sees us practicalise our newly forming theoretical knowledge through experimentation and action research in the classroom and institution. This practice then becomes part of our expanding experiential knowledge, especially as we continue to discuss and involve others in our research and experimentation. This continuous cycle from knowledge-seeker to knowledge-discusser and knowledge-user, where all three are often occurring simultaneously, results in increased understanding and growth in our professional knowledge.

Knowledge-provider

Eventually, as our knowledge grows, we begin to realise we have something to offer the professional community and begin to act as knowledge-providers.

The user, rather than the developer, takes on the role of knowledge-provider, and the exchange results in another user receiving the knowledge that he/she sought or to the website being expanded to provide the knowledge to all users. The exchanges under this role divided into the following turns; (developer request) – (user intention) – (developer acceptance) – user provision – developer or user thanks. The exchanges were as short as two turns with user provision and thanks, and as long as the full set of five turns. The example below was initiated by an open request for contributions to the site, and the resulting exchange provided examples of all the turns found in this role.

> Would you be willing to send me any of your assignments and have them posted?
>> (Turn 1, Developer to User 1.09, 27 January 2002)

> I don't mind sending some assignments to your site.
>> (Turn 2, from User 1.09, 27 January 2002)

> Attached is a TDA [MSc Text and Discourse Module] assignment and a 'task' to go with it. Feel free to add it/them to your website if you wish.
>> (Turn 3, from User 1.09, February 2002)

> I can't actually post assignments... However, if you would allow me, can I have a deeper look and see if it could possibly be reworked a bit to act as a stand alone page? Would you be OK with that? Let me know and thanks a million for your support.
>> (Turn 4, Developer to User 1.09, February 2002)

> Please can you check out your article via the newsletter link... Let me know if you want anything changed. http://www.philseflsupport. com/newsletter3.asp
> Thanks for your input and all your support as always.
>> (Turn 5, Developer to User 1.09, 1 October 2002)

So, the knowledge-provider role is defined as those users who are providing knowledge, with exchanges moving through a two- to five-turn discourse pattern, which included the turns of knowledge provision and thanks in every exchange.

However, it appeared that some users needed encouragement to gain the confidence to move onto this phase and share their knowledge with others, so many article contributions were solicited by the developer. Users who had for the most part never had their work published were freely willing to offer resources they had found to other users. They offered useful links to be uploaded to the site and, once the discussion boards were available, were prepared to offer advice and tips to fellow participants. However, not one of them freely offered one of their articles for inclusion unless they were solicited by the developer. When I knew a user's research from our email exchanges, I often requested a contribution from them and all those approached expressed their surprise and delight. I would suggest that this is indicative of how we all felt when we first started in the profession and needed encouragement from a more experienced colleague to give us the confidence to submit our first article for publication be it in a journal or on the web (see Wharton, this volume).

This role is the final phase in the cyclical model of teacher knowledge and is the key reporting element, which often strengthens our knowledge structure and teacher beliefs, perhaps moving them to the central core of our professional selves. However, the cycle does not end here, since a natural continuation of knowledge provision is a realisation of how much more we can add to our knowledge, so we begin to seek new knowledge in a process of lifelong learning. The cycle continues and grows.

The knowledge cycle

The four roles together act as a description of the full model of teacher knowledge development, bringing out the continual cycle created through the theorising of practical knowledge and the practicalising of theoretical knowledge.

In effect teachers are knowledge-seekers, knowledge-discussers, knowledge-users and knowledge-providers every day of their professional lives. Knowledge is never static as we continually collaborate, research and work towards extending and deepening our understanding of the profession and the subject we teach.

So, we now see a model which reflects an ever-widening cycle of development, with the knowledge of self at the centre upon which everything else is based. This knowledge of self ties together our informal, experiential and practical knowledge from context and from the theoretical formal knowledge of our profession, which we continually seek to extend by theorizing our practice and practicalising our theory. The outer circle demonstrates how this works in practice with teachers continually seeking knowledge to better understand their classroom practice, and then vocalising the reflection on that knowledge through discussion whilst using it in practice. The final stage sees the confident teaching professional providing their knowledge to others whilst they continue to better understand their own practice in an ever-widening cycle of teacher knowledge development.

The theoretical model of teacher knowledge that I have presented in this chapter is an attempt to provide a diagrammatical representation of how we construct and process knowledge. It addresses the 'tighter coupling of theory and practice in the context of a broader and deeper base of knowledge about learning, development and teaching' which Darling-Hammond saw as 'the key feature of teacher education for the twenty-first century' (1999: 227). And finally, it meets Johnson's definition of professional development as 'a collaborative effort, a reflective process, a situated experience and, a theorizing opportunity' (2002:1).

The model brings together as a coherent whole the 'formal theoretical knowledge' that is 'publicly represented' and 'negotiable' (Bereiter and Scardamalia, 1993: 62) and delivered through formal education programmes such as Diplomas, Degrees and Masters with the informal practical knowledge that is personalised, situated in context and, more often than not, learnt on the job.

Overall, the website has proved to be a success as a support site for teachers studying for further diplomas or Masters. The response from other users including teachers far removed from TESOL indicates that the site has fulfilled a need for teachers, especially those who rely on the web whilst studying at a distance.

Practical advice

Feedback from website users indicated that the nature of the developer discourse was essential to their continued involvement in the site. This discourse strategy can be defined by nine key elements which are relevant to both site users and site developers:

Key 1   Respond rapidly

This key aimed to avoid the experiences of other developers who had found that email exchanges often failed to move beyond the initial contact because it took too long to obtain a response (Kamhi-Stein 2000). I attempted to respond to all emails within 24 hours and I would also encourage all teachers using the web to develop a similar strategy in building strong online collaborative partners for their research.

Key 2   Show interest

The literature on computer-mediated communication (Salmon 2000) and journal writing (Mlynarczyk 1998; Quirke 2001a) stresses the importance of building trust, sociability, emotional support and a sense of belonging, all of which are reached primarily by showing a genuine interest in the people with whom you are communicating.

Key 3   Maintain a 'friendly' mood

The literature emphasises the importance of the leader's role and attitude in computer-mediated communication (Hyde 2000). In my website developer role, therefore, I attempted to maintain an upbeat, optimistic and friendly mood. This was done in many ways, for example through the use of smiley faces (☺) and colloquial terms. The friendly mood maintained by the language in both the emails and on the site was the feature most commented on by users, and I would advise all teachers to adopt this key.

Key 4   Respond in detail

One of the main methods of providing support to teachers studying at a distance is to try and give them the detail which they are looking for in their knowledge search. In my detailed responses, I aimed to give the knowledge-seekers a personal description of my experience (Hammond 2000) and understanding as well as links to other resources. In this way, the users were provided with a narrative and sources they could reflect upon (Schön 1983; Wallace 1991) to further develop their research in line with their teacher development beliefs (Tsui 2003) that underpin the model presented in this chapter. Teachers seeking knowledge can take full advantage of this key as the more detailed they are in their requests and responses, the greater the interest they will generate in their prospective collaborator.

Key 5   Empathise

The literature indicates that empathy is central in professional development relationships (Golombek and Johnson 2004) and, wherever possible, I related the user's difficulties or situation to my experience, and this was reflected in the feedback from the users, with one user stating that 'Phil tried to build a feeling of sitting in the same boat', and another concluding: 'I especially appreciate that Phil...lends a sympathetic ear.' It should be noted that the literature does not restrict the centrality of empathy to developers or knowledge providers but extends this key to all of us building professional development relationships.

Key 6   Ask frequent questions

I have long experienced the power of questions in developing communication through my work on journals (Quirke 2001a), and the literature on teacher development has numerous references on how questions can best be used to promote deeper reflection (Williams and Burden 1997). Questions were used to encourage response so that the discourse and contact could continue and allow a closer relationship between the user and the developer to be built. By having every email response include at least one direct question, I attempted to ensure that the emphasis was on the users continuing or ending the exchange. Of course, the most adept online communicators seeking knowledge used exactly the same strategy by continually using questions for me to respond to, placing that emphasis on me.

Key 7   Encourage contributions and participation

The website was set up to support teachers, and in order for this support to remain true to the collaborative and collegial nature of teacher knowledge development (Woods 1996), it was essential to openly and continuously call on the contributions and participation of the users. I attempted to do this through the language of the website, the use of discussion boards and continual encouragement in the email exchanges for users to contribute. Users commented on how these requests to contribute boosted their morale and gave them a real sense of ability (1.06), and 'how the mixture of site and personal email encouraged contribution' (User 1.07). This boost in confidence ties into the model of teacher knowledge development as users were encouraged to share their work and research as knowledge-providers. This acceptance into the profession as equals by someone they saw as knowledgeable clearly boosted confidence and encouraged further development. I also witnessed some users calling on me to contribute to their websites and research with a direct request to participate in their work – a very effective strategy I found as I felt encouraged and boosted by their confidence in me.

Key 8   Solicit comments and opinions

The above key was supported by the solicitation of comments and opinions from the users. These were an integral part of the developer discourse strategy in almost every email sent since the whole goal of the website as the research vehicle was to allow it to develop based upon user needs. This key aimed to ensure that I would receive regular feedback and suggestions. I am convinced of the effectiveness of involving all participants in planning decisions through feedback (Head and Taylor 1997, Quirke 2001b), and I continued this belief into my website development work. This key was a direct result of that conviction and the volume of feedback received was an indication of the success of this approach.

Key 9   Express support

This final key was central to the whole discourse strategy and the foundation guiding the other eight keys. As the aim of the website was to support teachers doing further studies, every developer email attempted to stress the importance of the support the site was trying to give.

The keys above indicate a discourse strategy based on the principles of users as co-developers and the developer as a supportive resource. This developer discourse was described by one user (1.10) as giving 'the sense of having a conversation' with 'the give and take' that that entails. The feedback from the users indicates that this discourse strategy appeared to meet its aim of supporting the users through the constant application of these nine keys to effective web facilitation. And I would also suggest that should knowledge seekers adopt the same keys, they can also build more effective and longer lasting collaborative relationships to support their studies and research.

For those who are interested in developing their own educational support site, the following statements of practical advice are suggested.

Statement 1

Build, develop and extend the site based upon input from the users to ensure that your website meets their knowledge-seeker needs.

Statement 2

Ensure the website asks questions, sets tasks to be completed and challenges the users of the site to ensure the site is fulfilling the aim of promoting reflection throughout the model of teacher knowledge development.

Statement 3

Give the users of the website as many avenues as possible to collaborate and interact with the site, the developer and other users through the full use of discussion boards and chat room facilities as well as regular emails. In this way you can give those users feeling isolated a sense of support and an avenue to discuss their developing knowledge. Thereby, you can meet the knowledge-discusser needs of users even though many of them will, in all probability, find other avenues for discussion away from the web and within their own context.

Statement 4

Build links into the website which give users access to forums where they can put their newly acquired knowledge into practice. These links could include sites such as web-based learning platforms like WebCT, survey sites like Survey Monkey (http://www.surveymonkey.com) and specific study sites like the AEROL site on action research (http://www.scu.edu.au/schools/gcm/ar/areol/areolind.html). In this way your website can act as an avenue towards offering connections to support the knowledge-user needs of the teachers who access the site.

Statement 5

Encourage users of the website to contribute their research, papers and potential publications, so that the website can act as a potential outlet for the knowledge-provider needs of the users as they grow in confidence.

Statement 6

Maintain the cycle of knowledge development by keeping in close and frequent contact with the users of the website and employing all nine keys to effective web facilitation to aid the teachers who use the site to meet the final principle of knowledge development.

Future developments – where do I go from here?

Recent huge advances in technology have created the possibility for educators to employ a wide range of new social constructivist techniques (Duguid and Brown 2001). As Stevens points out: 'Eventually chat tools will be integrated with daily life, as in the telephone, but at the moment their use in education is misunderstood, their potential as powerful learning tools nowhere near fully realised' (2005: 8), and he goes on to list the following tools:

Asynchronous CMC (computer-mediated communication) Tools
Yahoo Groups: http://www.yahoogroups.com
Moodle: http://www.moodle.com

Blackboard: http://www.blackboard.com
Nicenet: http://www.nicenet.net
Global Educator's Network: http://vu.cs.sfu.ca/vu/tlnce/PublicReg/PR_
Register.cgi
Wimba: http://www.wimba.com
Blogs
Wikki

Synchronous CMC (computer-mediated communication) Tools
Tapped-In: http://www.tappedin.org
Yahoo Messenger: http://messenger.yahoo.com/messenger/download/
index.html,
and other instant messengers:
Netmeeting: http://microsoft.com/windows/netmeeting
iVisit: http://www.ivisit.com
Pal Talk: http://www.paltalk.com
Alado voice portal: http://www.alado.net
Learning Times: http://www.learningtimes.net

I am now aiming to build upon my own research by developing a new diploma for English language teachers based upon the model. This diploma and its success will provide further validity claims, or not, for the model's robustness. During the diploma's first year, I will focus my attention on the role of reflection and knowledge processing and how they can be represented more clearly within the model. I envisage that this diploma could potentially demonstrate that the model has sound practical applications for the development of teacher education courses that match what we have learnt in the last few years about how we acquire knowledge.

Reflection is clearly a core characteristic of the development of teacher knowledge. The study has indicated that this is essentially a private, individual activity which can only be researched once it has been vocalised through knowledge discussion or the knowledge provision of personal narratives and case studies. I, therefore, feel it is important to include reflection explicitly within the model as a separate role and this will be one of my main focal points as I develop the new teacher development diploma.

Another future development I foresee is the launch of further teacher support sites using the statements of practical advice listed above. Such sites could also provide the potential for further study on, for example, how crucial the website developer role is, the petering out of CMC discussions, the importance of the length of turns in the knowledge-discusser role and the human element in web-based communication.

A final avenue of future development is the discovery of which technology tools could greatly enhance any site's ability to better address the knowledge-discusser role. This advancing technology will also allow new sites and further research to explore areas such as the reasons why teachers do not self-select onto potentially valuable resources and why lurkers do not get involved in, but continually return to sites. Another issue which we are likely to understand far better as technology advances is that of identity online and how our web-based communication develops as our online identity matures. This, in turn, may inform our understanding of how online discourse communities are created and the role they play in successful professional discussions.

I would like to conclude with the words of the users who contributed so much to this research and to whom a simple 'thanks' will never be enough.

> I think there are a lot of other teachers like me who are seeking to improve their credentials with a master's degree and are in need of guidance and support. You've found a need and are in the privileged position of filling it. That's great.
>
> (User 1.29)

Long may I be able to fill that need. There is nothing more professionally fulfilling.

References

Bereiter, C. and Scardamalia, D. (1993). *Surpassing Ourselves: An Inquiry into the Nature and Implications of Expertise*. Illinois: Open Court.

Brown, G. and Yule, G. (1983). *Discourse Analysis*. Cambridge: Cambridge University Press.

Darling-Hammond, L. (1999). 'Educating teachers for the next century: Rethinking practice and policy'. In G. Griffin (ed.), *The Education of Teachers*. Chicago: University of Chicago Press, pp. 221–56.

Duguid, P., and Brown, J. S. (2000). *The Social Life of Information*. Cambridge, MA: Harvard University Press.

Edge, J. (1992). *Cooperative Development: Professional Self-Development through Cooperation with Colleagues*. Harlow: Longman.

Golombek, P. R. and Johnson, K. E. (2004). 'Narrative inquiry as a mediational space: Examining emotional and cognitive dissonance in second-language teachers' development'. *Teachers and Teaching: Theory and Practice*, 10: 307–28.

Hammond, M. (2000). 'Communication within on-line forums: The opportunities, the constraints and the value of a communicative approach'. *Computers & Education*, 35: 251–62.

Head, K. and Taylor, P. (1997). *Readings in Teacher Development*. London: Heinemann.

Hyde, P. (2000). 'Towards a virtual learning community: Building a professional development website for the AMEP'. *Prospect*, 15(3): 65–80.

Johnson, K. (2002). 'Second language teacher education'. *TESOL Matters*, 12(2): 1 and 8.

Johnson, K. (ed.) (2000). *Teacher Education*. Alexandria, VA: TESOL.

Kamhi-Stein, L. D. (2000). 'Looking to the future of TESOL teacher education: Web-based bulletin board discussions in a methods course'. *TESOL Quarterly*, 34(3): 423–55.

McCarthy, M. (1991). *Discourse Analysis for Language Teachers*. Cambridge: Cambridge University Press.

Mlynarczyk, R. (1998). *Conversations of the Mind: The Uses of Journal Writing for Second-Language Learners*. Mahwah, NJ: Erlbaum.

Quirke, P. (2001a). 'Maximising student writing and minimising teacher correction'. In J. Burton and M. Carroll (eds), *Journal Writing: Case Studies in TESOL Practices Series*. Alexandria, VA: TESOL.

Quirke, P. (2001b). 'Hearing voices: A reliable and flexible framework for gathering and using student feedback'. In J. Edge (ed.), *Action Research: Case Studies in TESOL Practices Series*. Alexandria, VA: TESOL.

Salmon, G. (2000). *E-Moderating: The Key to Teaching and Learning Online*. London: Kogan Page.

Schön, D. A. (1983) *The Reflective Practitioner: How Professionals Think in Action*. London: Temple Smith.

Sinclair, J. and Coulthard, R. (1975). *Towards an Analysis of Discourse: The English Used by Teachers and Pupils*. Oxford: Oxford University Press.

Stevens, V. (2005). 'Creating online communities'. *TESOLArabia Perspectives*, 12(2), January: 6–11.

Tsui, A. B. M. (2003). *Understanding Expertise in Teaching*. Cambridge: Cambridge University Press.

Wallace, M. J. (1991). *Training Foreign Language Teachers: A Reflective Approach*. Cambridge: Cambridge University Press.

Wallace, M. J. (1997). *Action Research for Language Teachers*. Cambridge: Cambridge University Press.

Williams, M. and Burden, R. L. (1997). *Psychology for Language Teachers: A Social Constructivist Approach*. Cambridge: Cambridge University Press.

Woods, D. (1996). *Teacher Cognition in Language Teaching: Beliefs, Decision-Making and Classroom Practice*. Cambridge: Cambridge University Press.

8
'Metaphors keep cropping up': Dialogic Features of Metaphor in Exploratory Research Talk

Steve Mann

Introduction

This chapter presents an analysis of dialogic features of a group of teachers' exploratory talk. Its aim is to provide an illuminative account of two related processes evident in this distinct professional discourse. The first process is at the level of individual understanding, where teachers articulate ideas, thoughts and concepts. The second process is at the level of group understanding, where the nature of the group's discourse is considered and commented on. This shared understanding develops over time. The chapter argues that both these processes are dialogic and that metaphor plays a key role in each process. Data extracts are provided to demonstrate how metaphor plays this important dialogic role.

The author takes the view that teachers, at least occasionally, get the chance to talk together and it may be worth investigating better (or at least different) ways of talking in order to move professional development forward. The chapter shows how this might be done by illustrating how a different way of talking sets up different possibilities for dialogue. It provides concrete examples of how extra space allows exploration through extended metaphor. The chapter begins by describing features of this distinct discourse. It then outlines the relationship between articulation and four related dialogic processes. Having established this 'dialogic context', the particular dialogic role of metaphor will be examined.

Developing a distinct discourse

The first section of this chapter details features of a discourse developed by a group of six language teacher/teacher-educators. This detail enables the reader to understand the context in which the dialogic processes are evident. The featured group decided it would be useful to create a space during the week for a different kind of meeting that focused solely on the development of one individual member's ideas. While the group was committed to the idea that 'professional development takes place through professional conversations' (Crookes 1997: 68), it also felt that it was worth committing to a non-evaluative way of talking which might create extra space for articulation of ideas. The group began with the model of cooperative development (CD), as outlined in Edge (1992). The process of the group's use and development of 'cooperative development' is well documented elsewhere (e.g., Edge 2002, Mann 2002a, Mann 2002b). However, it is worth providing a short summary of its main features in order to understand the dialogic processes later highlighted.

In consciously developing a new kind of talk, we wanted the group to focus on one individual at a time (the 'Speaker'). The rest of the group acted as 'Understanders', giving the Speaker the benefit of the group's focused attention. The CD meeting involved between four and seven colleagues.

Distinct roles

Within this discourse, then, there are two roles (Speaker and Understander). The Understanders play a supporting role and help the Speaker to work on the development of ideas in a way that is not possible in normal talk or 'agenda driven meetings' (Mann 2004). The following section provides detail about these roles.

The Speaker

There is one Speaker in each meeting. Individuals within the group take it in turns to be Speaker. Being a Speaker is an opportunity to talk out an idea, an issue or a personal concern. The choice of topic is determined solely by the featured individual and may or may not have immediate relevance to the group. It is better if the Speaker has not planned the talk as it is important that this idea is emergent. It is not a space for 'reporting' ideas that have been previously polished, presented or published.

The Understander

Understanders consciously use CD's non-judgemental moves, particularly those termed *Reflecting, Focusing* and *Relating* (Capital letters are being used for both the roles and moves, in order to accentuate the specific nature of the terms.) In CD, the Understander tries to keep all aspects of evaluation out of their contributions and resist the temptation to offer suggestions and advice.

Distinct moves

Mann (2002a) in a two-year study confirmed that three 'discourse moves' were particularly important in the development of the group's distinct discourse. These are Reflection, Focusing and Relating (see Table 8.1).

Mann (2002a) established that Reflection can be considered the core move as elements of reflection are usually present in both Focusing and Relating moves.

Distinct phases

The group's CD sessions were approximately one hour. A session is divided into three distinct phases (see Table 8.2).

Individual members of the CD group were overwhelmingly positive in their comments on the value of their sessions as Speakers (see Edge 2002 and Mann 2002a). Interviews with participants established that they felt

Table 8.1 Discourse moves in CD sessions

Reflection

Giving back a version of what the Speaker has just said. It does not have to be word for word. It is not a case of 'parroting' the last thing said. Rather, the Understander is honestly trying to Reflect back a version of what has just been said.

Focusing

Offering something that the Speaker has previously said as a possible topic for further articulation. It might go something like 'a few minutes ago you said X, would you like to say a little more about that?' This kind of move can also be used early on in a session to establish the scope of the topic, idea or theme being articulated.

Relating

Taking two or more aspects of the Speaker's previous talk and presenting them back. It is often a case of saying 'you've said A and you've said B, how are they related?'

Table 8.2   Phases in CD sessions

Speaker stage	The Speaker speaks for 25–35 minutes with others acting as Understanders. During this time, the Speaker works on an emerging idea. The Understanders try to support the Speaker by either Reflecting, Focusing or Relating.
Resonance stage	The Understanders become Speakers and take turns to articulate a non-evaluative response (what the group came to call a 'Resonance'). The orientation of these resonances is much more of a 'what Harry has helped me to see is…' rather than 'I think I see this differently from Harry…'.
Concluding phase	After Understanders have shared their Resonances, the Speaker responds to these resonances and makes some concluding comment about what they have got out of the session or perhaps what the next step might involve.

they had 'extra space' to develop ideas and that there was 'a feeling of energy' and 'attention being concentrated' on a developing idea.

Initially this 'group development' process was for our group only. However, after two years we began to offer this possibility to a series of visitors. All of these visitors reported favourably on both the process and its outcomes. There were comments such as 'unique kind of support offered', 'rare opportunity' and 'a feeling of concentration and focus'.

From the discourse of debate to a dialogic discourse

The group's decision to adopt and develop a more non-evaluative discourse was a conscious attempt to open up new dialogic possibilities. It certainly was not intended to replace the cut and thrust of academic argument and informed debate. However, it was an attempt to open up a different dialogic space. The potential and uses of dialogic rather than debate discourse have been discussed by a number of writers (e.g., Barnett-Pearce and Pearce 2004; Pearce 2002 and Yankelovich 1999).

At this point it is necessary to say a little more about the scope of the term 'dialogic' used in this chapter. The term dialogic encompasses dialogue 'between' at different levels. The most obvious level is our commonsense notion of dialogue (talk between people). However the term also usefully covers an individual's *internal* dialogue. This individual level might include both unconscious and conscious thought and also verbal attempts to capture or further this thought. Such dialogue might be between aspects of the individual's identity (e.g., dialogue between a teacher's professional and personal 'self') or might arise out of a dialogue between competing work pressures or trying to achieve life/work balance (Spencer 1986). It might involve becoming more centred and connected (Palmer 1999) or trying to find 'somewhere to stand' (Clarke 2003). As we will see later in this chapter, metaphors of balance, footing and equilibrium are particularly significant in individual attempts to explore difficulties.

The term articulation is deliberately used here to accentuate the distinct and formative nature of the talk highlighted. Taylor uses the term 'articulation' to distinguish a particular kind of talk (distinguished from describing or informing). Perhaps the closest synonym would be 'formulation' (as in 'formulating her ideas'), where what is 'initially inchoate, or confused, or badly formulated' (Taylor 1985: 36) is worked on and improved. In the extracts of data that follow in subsequent sections, the intention is to demonstrate articulation and its relation to four individual and group dialogic processes:

1. The Speaker articulates something. On hearing this 'version', the Speaker qualifies, revises or extends that version. The extra space afforded to the Speaker allows the Speaker to articulate, listen to and work 'live' on his or her own ideas. There is a dialogic relationship between what the Speaker thinks, says and then hears.
2. The Speaker articulates something. An Understander Reflects a version of what the Speaker has said. There is a dialogic relationship between the two versions.

In addition, I want to pay particular attention to the role played by metaphor:

1. A Speaker uses a metaphor to express an emergent idea. There is a dialogic relationship between the vehicle (the metaphor) and the target meaning.
2. The Understander or Speaker uses a metaphor to try and capture some aspect of the group's discourse. This is taken up by others and consequently a dialogic process is engaged.

To sum up then, the expressed purpose of the CD sessions is to create optimum conditions for the articulation of emergent understandings and real-time thinking. Interviews with participants confirmed the findings of the analysis of the transcripts of the sessions (Mann 2002a) that this way of talking allows greater space for exploration than is possible in other kinds of professional talk. The purpose of the rest of this chapter is to detail the ways in which the discourse encourages the range of related dialogic processes (1–4) distinguished above.

Internal dialogue

The first example highlighted shows the first two processes (internal dialogue and Speaker–Understander dialogue). It shows how the latter supports the former. In this session Nick (Understander) is engaged in supporting Vince as Speaker. He picks out the key elements and gives Vince a chance to 'hear back' a version of his emerging focus. Vince is engaged in an internal dialogue where he is articulating the balance between planning and more spontaneous ways of talking in the classroom.

Extract 1

N = Nick (Understander); V = Vince (Speaker)

001	V:	as soon as I enter into a
002		planning world (.) in terms of
003		talking (0.4) it seems to cause
004		some kind of stress
005	N:	Mmm=
006	V:	=which I- which I feel imposing on
007		me (.) and this imposition, (.)
008		this structure that I've pre-
009		planned, (0.4) I find is-
010		is a saddle (.) a chain (.)
011		something which inhibits me.
011	N:	so can we just clarify where we
012		are now?(.) you're now into (.)
013		what may not be a continuing topic
014		but the first area of topic focus
015		is what you're working on now and
016		that is this preference of yours
017		for off-the-cuff talk as opposed
018		to planned talk. (.) you're saying
019		(.) that if you plan something
020		then when you start to talk (.)
021		you feel that that plan is an

022	imposition on you and constrains
023	you and ties you down and you feel
024	you're not being as productive as
025	you could be.

Vince is articulating an internal dialogue. In this case his 'received knowledge' (he should plan lessons) is in conflict with his 'experiential knowledge' (Wallace 1991:15) and Nick's Understander move helps Vince explore the extent to which his experience suggests a more 'improvised' and more spontaneous way of working may more comfortable for him. In what follows, it is worth noting how the metaphor of the 'athlete' is consciously used to further the emerging preparation/planning distinction:

Extract 2

V = Vince (Speaker)

001	V:	yes yes (.) and another thought
002		hits me from that (.) it's- it's
003		the preparation planning
004		distinction (.) that an athlete
005		doesn't necessarily prepare for
006		the 100 meters by doing the 100
007		meters (.) they prepare in lots of
008		different ways (.) so planning for
009		a speech event (.) if you take
010		that metaphor to its conclusion
011		(.) it's not a good way (.) to-
012		(.) to prepare (.) by preparing in
013		exactly the same way as the speech
014		event is going to- (.) to are take
015		the form of (.) there are probably
016		other ways which the limbering
017		up (.) the warming up exercises...

This is an interactive and dialogic process in the pursuit of greater understanding. At its best the concentration on the Speaker's ideas produces thinking in real time. Elements of new thought, speculation, discovery and realisation are often lexically signalled (e.g. 'another thought hits me' in lines 1–2 above).

In many of the CD sessions the Speaker is bringing to the meeting something that they feel is causing some difficulty or something they feel they need to make progress on. By the end of the session there is often an expression of relief that something has been resolved or moved forward. The following extract is taken from the end of a meeting where Harry has been the Speaker talking about a research idea. There is a strong sense that it has been valuable and that the purpose has been achieved. He has brought a topic to the meeting that has been problematic and he feels that progress has been made:

Extract 3

H = Harry (Speaker)

001	H:	I really wanted to:: >you know< to
002		try and bring up something in the
003		research that was not a bl<u>ock</u> but
004		was actually causing me pr<u>o</u>blems
005		(.) which I <u>have</u> done (.) and as a
006		result I think I've got a clearer
010		p<u>i</u>cture (.) of some of the w<u>o</u>rries
011		and where they're <u>c</u>oming from and
012		what I can- what I should be a<u>w</u>are
013		of (.) and what I can afford to
014		just leave on the back burner (.)
015		and that's helped me (.) I think
016		I'm more in b<u>a</u>lance about the way
017		that I'm going to go a<u>bout</u> this
018		now…

We get a clear sense that worries and problems have been addressed, and for later purposes, we might also note Harry's use of metaphors 'a clearer picture', 'on the back burner' and 'more in balance'.

Speaker–Understander dialogue

We need to consider two dialogic relationships between the Speaker and the Understanders. The first occurs in the Speaker stage where the Understanders are using the distinct discourse moves of Reflecting, Focusing or Relating to support the Speaker and create the space for further internal dialogue. We have already seen Nick doing this in Extract 1 above. The second occurs in the Resonance stage where the Speaker responds to the Understanders' Resonance statements.

We can see the first of these dialogic processes in Extract 4. In this session Elizabeth is talking about her research into the recognition of lexical chunks. Vince picks up three elements of what Elizabeth has been talking about (lexical chunks can either be identified by intuition, corpus analysis or phonologically). Elizabeth has spoken at length about identification through intuition and through corpus analysis. However, despite a previous mention of intonation, Elizabeth has not spoken about phonological identification. Vince's move has elements of Reflection and Focusing. Its function is to ask Elizabeth whether it would be useful for her to talk about this third possibility:

Extract 4

V = Vince (Understander)

001	V:	I was just interested (.) you
002		talked- you talked about it- you
003		concentrated on intuition and (.)
004		statistical corpus analysis but
005		you also mentioned a third way of-
006		of recognising chunks which was
007		intonation (.) but you didn't talk
008		about that as much- you didn't
009		talk about that as much (.) you
010		didn't pick that up (.) do you see
011		that as a future (.) something (.)
012		that teachers can work with in the
013		classroom?

In Elizabeth's subsequent turn we can see that this is not something that she has previously thought about. We can see elements of being unsure. This is real-time thinking and the Understander–Speaker dialogue (between Vince and Elizabeth) has created space for further internal dialogue. After saying what she has not done, she begins to speculate:

Extract 5

E = Elizabeth (Speaker)

001	E:	urhmm (.) .hhh I haven't
002		researched (.) ummm (.) I don't
003		know of any research that does and
004		I haven't done any research into

005	the phonological aspects (.) of
006	(.) urrhmm (.) I would have
007	thought that they'd be an awful
008	lot of chunks (.) phonological
009	chunks that are not necessarily
010	(.) ummm highly frequent (.) ummm
011	highly frequent pre- fabricated
012	chunks (.) for one thing is they
013	might be just be highly frequent
014	for that person because they're
015	into that topic or into that
016	subject (.) I don't know (.)<I
017	mean>I have absolutely no idea (,)
018	would imagine (.) that we'd need
019	because I quite a lot of research
020	to find out <I mean> now we've got
021	the frequency lists (.) and we've
022	got the same for spoken as well
023	(.) it would be really interesting
024	to get onto do some phonological
025	research and actually see whether
026	the chunks that come up without
027	pause (.) are on this list (.) I
028	mean it could be done (.) yes I
029	hadn't thought of that

This is an internal dialogue between what has not been done (in the past) and what might be possible (in the future). Through the talk Elizabeth moves from a series of statements with negative polarity (haven't done/don't know) to more speculative and positive modality (would/could). The real-time cognitive process can be glimpsed in 'now I'm thinking' and 'I would imagine', 'yes, I hadn't thought of that'.

The dialogic role of metaphor

A number of writers have detailed the use of metaphor by individuals for reflective purposes (Block 1996; Bullough 1991; Bullough and Stokes 1994; Elliot 1994; Farrell 2006; Munby 1986; Thornbury 1991; Tobin 1990). It has been argued that individuals use metaphoric construction as an 'introspective and reflective tool... tapping the kinds of meanings practitioners create about their own professional actions, practices and personal theories' (Burns 1999: 147) and that metaphors can be used as 'explanatory vehicles' (Block 1996: 51–3). The exploration of ideas through metaphor is an integral part of reflective practitioner thought and, as Oxford and colleagues state, such thought and reflection is 'part of the ongoing life of each language teacher' (1998: 46).

The remainder of this chapter concentrates on the particular ways in which metaphor gets used both to articulate individual ideas and understand the nature of the distinct discourse being developed by the group.

Internal dialogue and metaphors of balance

Articulation helps the individual to get into a more centred or balanced state. In Extract 3 Harry talked about being 'more in balance'. In Extract 2 we saw how Vince used part of a session to articulate a balance between being planned and being prepared. Vince uses a metaphor (the metaphor of a 100 meters athlete) to sustain an internal dialogue between received knowledge and experiential knowledge (Wallace 1991: 15). Other internal dialogues might be between our personal and public theories (Griffiths and Tann 1992), private and professional selves, *ideal self* and *actual self* (Rulla, Imoda and Rideck 1978), or espoused theory and theory-in-action (Argyris 1980).

Another session reveals several examples of articulating this kind of balance. In Extract 6, Emma has been talking about her move from full-time to part-time employment, hoping to achieve a better life/work balance. She is reflecting on how this might change the nature of her work. In the Resonance stage, several Understanders refer to this personal balance:

Extract 6

V = Vince (Understander)

001	V:	and so (.) getting that balance
002		(.) right (.) and even if it's a
003		fairly pragmatic and best shot
004		balance (.) it- you need to
005		constantly (.) reassess and think
006		about it (.) and it's almost on an

007	everyday basis that you think
008	well:: (.) I've done all I can
009	do::, (.) in that (.) in my
010	professional self today (.) and
011	because you've reached that
012	feeling of-of (.) contentment or
013	fulfillment with that part of
014	yourself then (.) your parenting
015	self can do all the things with a
016	clear conscience and therefore do
017	them properly

At the end of the session, Emma provides an explicit reference to the dialogic process that can happen between the Understander's Resonances and the Speaker's reflection. In this example she comments on Harry's Resonance:

Extract 7

E = Emma (Speaker)

001	and then what you ((Harry)) said
002	about work and enjoyment (.) that
003	really made me think of one of my
004	important reasons for (.) being
005	part-time (.) which again (.) one
006	of big hopes (.) is that (.) when
007	I have (.) errm (.) a smaller
008	workload (.) I will again feel
009	free to really enjoy it (.)
010	because before I had Noah (.) when
011	I could expand into a weekend or
012	whatever (.)

Extended metaphor

One of the significant features of the way metaphor is used in this exploratory talk is the prevalence of what might be termed extended metaphors. Previous studies of metaphor in teacher talk and thinking have focused on metaphor at the clausal level (e.g., Block 1996; Bullough 1991; Farrell 2006; Munby 1986; Oxford, Tomlinson et al. 1998; Thornbury 1991; Tobin 1990). It could be argued that these studies present metaphor in relatively fixed ways, particularly those that are trying to establish 'root metaphors' (Massengil, Mahlios and Barry 2005; Oxford, Tomlinson et al. 1998) and that they ignore the more exploratory and extended role of metaphor. This section looks at several examples of such extended metaphors.

Carousel

In this session the Speaker (Ella) was a visiting Speaker talking about an emerging research issue. In line 1 below she introduces a personal metaphor of a 'carousel' to represent her research process:

Extract 8

```
E = Ella (Speaker)
001      E:      like a carousel (.) it's all going
002              round and the research question is
003              actually the pole in the middle,
004              and everything else (.) your data
005              collection, your analysis, (.)
006              somehow has to relate (.) somehow
007              (.) right from the horses on the
008              outside to the pole (.) in the
009              middle (.) it's almost as if it
010              all goes round and somehow you
011              pick up teacher intention on the
012              way (.) <you know?>
```

A few moments later Harry (Understander) begins to work with this metaphor. We can see how key elements of the metaphor ('horses', 'movement') are reflected. This is essentially a check on what Ella is saying ('so what **you're** saying is'/'if I've understood **you**'). However, Harry is inevitably adding his understanding in a conscious way ('if **we** carry on with this metaphor') in order to better understand how the metaphor is working for Ella:

Extract 9

```
H = Harry (Understander); E = Ella (Speaker)
001      H:      So what you're saying is (.) <if
002              I've understood you correctly> if
003              you watch- if we can carry on with
```

004		this metaphor (.) these horses
005		moving up and down (.) if you
006		watch a sufficient number of
007		horses and sufficient amounts of
008		movement (.) you'll understand
009		more about the mechanism that's
010		actually (.) driving it (.)
011		[ ]
012	E:	driving it=
013	H:	=driving it, yeah

It is noticeable that both Harry (Understander) and Ella (Speaker) are involved in sustaining and to some extent co-constructing this metaphor. The co-operative overlap in lines 11/12 is a clear indication of this co-construction. There are several more exchanges that help Ella develop the metaphor and its significance for her. Ella later reported that developing this metaphor had been 'tremendously useful' for her in her subsequent thinking.

Hot Pipes

The next example of extended metaphor is typical of where a narrative is used metaphorically to further an emerging idea. In this case Harry is talking about a trip to Epcot (Disneyland). He recounts one exhibit where holding on to two metal bars (one slightly cold and one slightly warm) tricks the mind into thinking the hands are being burnt. The following is part of a longer anecdote:

Extract 10

H = Harry (Understander)

001	H:	all that's in the middle are
002		these two bars (.) the cold ones
003		and the warm ones (.)but when you
004		put them together the body can't
005		distinguish (.) the senses CANnot
006		distinguish between- it confuses
007		the senses and the message it
008		gives is it's burning (.) and the

The metaphor is being used to explore the issue that sometimes thinking and feeling seem mutually incompatible and it takes a great effort of will and concentration to resolve them:

Extract 11

H = Harry (Understander)
```
001      H:      and the wonderful thing about it
002              (.) sometimes KNOWing it isn't
003              quite enough(.) that the visceral
004              thing is just sort of even more
005              powerful
```

Awareness of metaphor use

There is a great deal of evidence that, not only are metaphors used, but individuals are very conscious of their use, they are remembered over time and also referred to by individuals in the group. This awareness of metaphor use can be demonstrated in the following examples. Here Nick is revisiting a metaphor in an article he had published some years previously. There is reference to an old and individual metaphor being used in a different way:

Extract 12

N = Nick (Speaker)
```
001      N:      and:: I'm coming back around to
002              them now:: and now they're
003              starting to be meaningful to me
004              (.) clearly again only in
005              metaphorical terms (.) but (.) for
006              as long as the metaphor holds (.)
```

In the next example we see metaphor playing a dialogic role. Robert (Understander) is Reflecting elements of the metaphor ('bubbling up' and 'trace elements'). However, in trying to establish the nature of the metaphor, he has misunderstood its origin. This enables Harry to clarify with further metaphorical articulation:

Extract 13

R = Robert (Understander); H = Harry (Speaker)
```
001      R:      =it leads me to wonder whether the
002              bubbling up °which I understand°
003              (.) the bubbling up means it hits
```

004		the surface
005	H:	mmm
006	R:	right? (.) IF nothing - if there
007		is something that doesn't hit the
008		surface it's not for you a trace
009		element
010	H:	Oh WELL (.) NO because it's - I
011		use the geological metaphor where
012		you can find in >you know< (.) you
013		dig up the soil that you know is
014		(.) of a particular type (.) that
015		in it there are tiny other bits of
016		in the soil (.) like bits of LEAD

Clearly, the explicit use of the word 'metaphor' (e.g. line 11) confirms that metaphor is being consciously used. Interviews also confirmed the group was conscious of prominent metaphors in sessions. As well as playing a dialogic role for the individual Speaker, they seem to be memorable for Understanders too. Using previously introduced metaphors (Extracts 9 and 10) we can see this in the following interview comments

Extract 14

V = Vince		
001		I mean I remember that one about
002		Harry's hot pipes (.) do you
003		remember that one where he talked
004		about holding onto two pipes in –
005		in Disneyland

Elizabeth in an interview about the CD process raises the issue of metaphor and there is a sense of both the excitement and response to individual metaphors:

Extract 15

E = Elizabeth (Interviewee)		
001	E:	errm I want to start off with
002		something really present because
003		one thing that really excites me

004	i::s (.) one very small aspect (.)
005	is the aspect of metaphor (.) as
006	metaphors keep cropping up

In another comment on prominent metaphors, Elizabeth recalls the 'carousel' metaphor and says:

Extract 16

E = Elizabeth (Interviewee)

001	the metaphor of the carousel (.)
002	was really strong and that really
003	sparked off a huge lot of thought
004	in me

Metaphor and shared understanding of the discourse

One of the most interesting aspects of metaphor use is the way in which the group uses metaphors to develop group understanding of the exploratory talk itself. This section provides some examples of where metaphor is used to comment on the perceived process. Thornbury (1991) points out that metaphors such as 'language as matter' have been pervasive in talking about language learning (language can be 'chunked' and 'segmented' and in fluid form 'filtered' and 'blocked'). In a similar way, the group tries to find metaphorical constructs to represent the space and energy created by this exploratory talk. In the first example, Nick is talking about the role of the Understander. In this example, Nick is drawing on the term 'perturbation':

Extract 17

N = Nick

001	N:	and you're enabling me to:: (.)
002		keep my thinking in a state of
003		perturbation for longer than one
004		is normally allowed to(.) in
005		company (.)

Nick's use of the term 'perturbation' derives from Rogers (1980) and in turn from Prigogine (1979) (a state of unstability or fluidity) and this term becomes an important one of the group's discussions. Rogers felt that opportunities for perturbation can help the individual work on complexities and inconsistencies and work towards 'a new, altered state, *more* ordered and coherent than before' (1980: 131). Prigogine and Stengers (1985) also uses the term perturbation to suggest that individuals can work from 'a far-from-equilibrium position' to a more centred state.

As well as greater energy being produced, the group becomes conscious of the way that group understanding sometimes offers far more than individual understanding. In the following extract Robert compares what is possible on an individual thought level (flatland) with what seems to be possible with 'group understanding'. He develops this comparison with a 'holographic' metaphor. Again it is interesting to note Harry's support for the construction of the metaphor with cooperative overlap in lines 8 and 15:

Extract 18

R = Robert (Speaker); H = Harry (Understander)

001	R:	=if I could coin a phrase
002		holographic understanding(.) if I
003		look at my own thoughts (.) I see
004		flatland (1.4) but if you have
005		group understanding you see
006		holographically (.) you know
007		you ['re- (1.0) you- you::
008	H:	[>where you're coming from=
009	R:	=you see the same object
010	H:	>different< yeah
011	R:	you see the same object=
012	H:	=yes
013	R:	but you see it from- from (.)
014		you see different aspects=
015	H:	= you can actually get round
016	R:	yeah

Metaphors are also used to try to better understand the distinct roles and moves. One important metaphor for the group is of a 'mirror'. The mirror metaphor works for the group in different ways. Sometimes the Understanders are like mirrors. We Reflect back and this gives the Speaker the chance to take a break and check on how the articulation is progressing. Sometimes we see ourselves as significantly different to mirrors. In the following example Harry sees the mirror metaphor as a helpful goal in that we try to 'give it back exactly' but this is, at the same time, impossible because we are not mirrors:

Extract 19

```
H = Harry (Understander)
001      H:      no but I'm saying you should try
002              to avoid that (.) it comes and you
003              do try to give it back exactly as
004              it comes in but ine::vitably (.)
005              we're not mirrors (.) but what you
006              try to do is not put some of
007              yourself in
```

In another session Elizabeth uses the mirror metaphor to make the point that what the Understander Reflects is not necessarily the same as what the Speaker thinks s/he has said:

Extract 20

```
E = Elizabeth (Understander); H: = Harry (Understander)
001      E:      you have this image of how you
002              look (.) and when you look in the
003              mirror you think oh shit!
004      H:      ⌈yes exactly!
005      E:      ⌊oh god yeah (.) I
006              didn't realise I had grey hair (.)
007              I'd forgotten
008      H:      yeah
009      E:      so Reflecting back (.) is more of
010              a mirror image of Reflecting is
011              not (.) it's not an exact (.)
012              it's not exactly going to match
013              what was in the mind of the
014              Speaker
```

A few turns later Nick uses Elizabeth's metaphor of the face in the mirror to make a contrast. Unlike our faces, our ideas can be improved, we can develop them:

Extract 21

```
N = Nick (Understander)
001      N:      and the- the– the real payoff
002              comes at that point in this sort
```

003	of work (.) unlike the shape of my
004	nose >and the rest of it< (.) it
005	is my ideas that we're talking
006	about (.) I can in fact improve
007	them (.) I can work on them

Conclusion

Looking back at the development of this particular professional group, it is evident that not only did we create opportunities for working on, developing and improving ideas but that we began to seriously evaluate and understand the opportunities for talk that are available in our professional lives. About six months after we started this work, Deborah Tannen's book *The Argument Culture* appeared in which she says:

> We need to use our imaginations and ingenuity to find different ways to seek truth and gain knowledge, and add them to our arsenal – or, should I say, the ingredients for our stew.
>
> (1998: 298)

The group felt that establishing a new way of talking was a step forward in meeting this kind of challenge. The benefits in shaping group identity and encouraging support and communication were tangible.

This chapter has paid particular attention to the role of metaphor in the development of this discourse. Interviews suggest that the group feel that metaphors play an important role in understanding the nature of this exploratory talk. When the participants try to pin down the nature and importance of the group development process itself, it is often with the help of metaphor that these understandings are attempted.

In terms of the dialogic processes evident in the featured group's exploratory discourse, the use of metaphor may be more extended, memorable and explanatory than has previously been realised in the field of teacher development. It is clear from the data that there are numerous 'extended' metaphors (i.e., not relatively fixed lexical phrases at the clausal level). They are often relatively 'fresh' and are adopted as exploratory or explanatory vehicles trying to achieve a particular semantic purpose. In some cases metaphors are revisited (e.g. 'coming back round') and reinterpreted. It may well be that, when we are reaching for new meanings, resolving tensions, exploring contradictions and articulating emergent understandings, such extended metaphors play an important role in the articulation of individual ideas.

Acknowledgements

I should like to thank all those who participated in the CD group for their generosity in giving time for interviews, discussion and comment. A lot of what is contained in this chapter results from their generosity. All errors are theirs and not mine☺.

If you are interested in setting up or researching discourse choices in teacher meetings or exploratory talk, please contact Steve Mann at CELTE, University of Warwick. Email: steve.mann@warwick.ac.uk

References

Anderson, R., Baxter, L. A., and Cissna, K. N. (eds). 2004. *Dialogue: Theorizing Difference in Communication Studies*. Thousand Oaks, CA: Sage.

Argyris, C. 1980. *Inner Contradictions of Rigorous Research*. New York: Academic Press.

Barnett-Pearce, W. and Pearce, K. 2004. 'Taking a communication perspective on dialogue'. In Anderson et al. (eds), pp. 39–56.

Block, D. 1996. 'Metaphors we teach and learn by'. *Prospect*, 73: 42–55.

Bullough, R. V., and Stokes, D. K. 1994. 'Analyzing personal teaching metaphors in preservice teacher education as a means for encouraging professional development'. *American. Educational Research Journal*, 31(1): 197–224.

Bullough, R. V., Jr. 1991. 'Exploring personal teaching metaphors in preservice teacher education'. *Journal of Teacher Education*, 42(1): 42–51.

Burns, A. 1999. *Collaborative Action Research For English Language Teachers*. Cambridge: Cambridge University Press.

Clarke, M. A. 2003. *A Place To Stand: Essays For Educators In Troubled Times*. Ann Arbor, MI: University of Michigan Press.

Crookes, G. 1997. 'What influences how second and foreign language teachers teach?' *The Modern Language Journal*, 81(i): 67–79.

Edge, J. 1992. *Co-Operative Development: Professional Development Through Co-Operation With Colleagues*. Harlow: Longman.

Edge, J. 2002. *Continuing Cooperative Development*. Ann Arbor, MI: University of Michigan Press.

Elliot, R. K. 1984. 'Metaphor, imagination and conceptions of education'. In W. Taylor (ed.), *Metaphors of Education*. London: Heinemann Educational, pp. 38–53.

Farrell, T. S. C. 2006. '"The teacher is an octopus": uncovering preservice English language teachers' prior beliefs through metaphor analysis'. *RELC Journal*, 37: 236–48.

Griffiths, M. and Tann, S. 1991. 'Ripples in the reflection'. In P. Lomax (ed.), BERA Dialogues No. 5: *Managing Better Schools and Colleges: An Action Research Way*. Clevedon: Multilingual Matters.

Mann, S. 2002a. 'CD: Cooperative development or continuing difficulties'. In J. Edge (ed.), pp. 218–24.

Mann, S. 2002b. 'Talking ourselves into understanding'. In K. Johnson and P. Golombek (eds) *Teachers' Narrative Inquiry as Professional Development*. Cambridge: Cambridge University Press, pp. 159–209.

Mann, S. 2004. 'Dialogic understanding'. *JALT 2003 Conference Proceedings*. Tokyo: JALT, pp. 116–24.

Massengill, D., Mahlios, M. and Barry, A. 2005. 'Metaphors and sense of teaching: How these constructs influence novice teachers'. *Teaching Education*, 16(3): 213–29.

Munby, H. 1986. 'Metaphor in the thinking of teachers: An exploratory study'. *Journal of Curriculum Studies*, 18(2): 197–209.

Oxford, R., Tomlinson, S., Barcelos, A., Harrington, C., Lavine, R. Z., Saleh, A. and Longhini, A. 1998. 'Clashing metaphors about classroom teachers: Toward a systematic typology for the language teaching field'. *System*, 26: 3–50.

Palmer, P. J. 1999. *The Courage To Teach: Exploring The Inner Landscape Of A Teacher's Life*. San Francisco: Jossey-Bass.

Pearce, K. A. 2002. *Making Better Social Worlds: Engaging in and Facilitating Dialogic Communication*. Redwood City, CA: Pearce Associates.

Prigogine, I. 1979. *From Being to Becoming*. San Francisco: W. H. Freeman.

Prigogine, I. and Stengers, I. 1985. *Out of Order Chaos*. London: Flamingo.

Rogers, C. R. 1980. *A Way of Being*. Boston: Houghton Mifflin.

Rulla, L., Imoda, F. and Ridick, J. 1978. *Psychological Structure and Vocation*. Dublin: Villa Books.

Spencer, D. A. 1986. *Contemporary Women Teachers: Balancing School and Home*. New York: Longmans.

Tannen, D. 1998. *The Argument Culture*. London: Virago.

Taylor, C. 1985. Human Agency and Language. Cambridge: Cambridge University Press.

Thornbury, S. 1991. 'Metaphors we work by: EFL and its metaphors'. *ELT Journal*, 45(3): 193–200.

Tobin, K. 1990. 'Changing metaphors and beliefs: A master switch for teaching?' *Theory into Practice*, 29(2): 122–7.

Wallace, M. 1991. *Training Foreign Language Teachers: A Reflective Approach*. Cambridge: Cambridge University Press.

Yankelovich, D. 1999. *The Magic of Dialogue: Transforming Conflict into Cooperation*. New York: Simon & Schuster.

9
Making the Break: Establishing a New School

Keith Richards

Introduction

> having been a few times round the block, teachers may be ready for new challenges, for new stimulation.
>
> (Huberman 1992: 124–5)

Huberman's observation on teachers with 7-18 years of experience in the classroom highlights a key moment in many careers. The instinct for discovery or adventure that attracts people to language teaching sooner or later leads them to unexpected places. The pursuit of greater understanding leads some into higher education, while a desire for greater independence or control over their professional circumstances prompts others to dip their toes in the uncertain waters of freelance activity. Occasionally, if the market permits and sufficient money can be scraped together, some teachers take the more drastic step of going into business for themselves, establishing their own language school and shifting their professional life onto a precarious balance of risk and reward. What follows is a case study of such a move, following its complete trajectory from conception to conclusion.

To my knowledge, there are to date no similar studies of these moments of profound change in the ELT professional life cycle, and there seems to be no obvious equivalent in mainstream education. This story is therefore unusual and I shall try to draw lessons from it, but what follows may also be valuable as a case study of a particular school, contributing to what Stenhouse (1980: 5) once described as 'the archaeology of the future'.

The story itself will be told, as far as possible, in the words of those who were involved in it and I shall limit my contributions to some necessary linking and the occasional highlighting of emergent issues. But in addition to the teachers' accounts of their experiences, I shall include some extracts from their talk because it seems clear to me that their journey to professional independence was also a *linguistic* journey and that awareness of this dimension can make an important and enriching contribution to our understanding of the professional world of TESOL.

A background to the case

The Pen school is in many ways fairly typical of small independent language schools the world over and the words of one of its teachers (Paul) will find an echo in many staffrooms: 'We are a small, struggling school who've built up a faithful following.' In common with other schools of its kind, it has a dedicated core of experienced professionals, takes in temporary teachers as needed, depends largely on its reputation in order to maintain student numbers, is open more or less throughout the year, has a particularly busy summer period, and is flexible in terms of what it offers.

Paul is one of the five core teachers who, by the time I joined them as a researcher, had worked together for between 11 and 18 years. Jenny is the principal and the other teachers form working pairs: Harry and Paul are responsible for general English courses, while Annette and Louise deal with business courses. I spent 15 months in the school as a participant (teacher) observer, attending one full day a week and for longer periods in the summer, producing a total of 57 days' attendance in all. The case that follows draws on a dataset comprising fieldnotes and interviews, as well as transcriptions of staff meetings and every 20-minute morning break throughout a single term. It has its roots long before the foundation of the Pen, in the participants' first experiences of teaching.

Drifting into a career

> I suppose we all sort of drifted into EFL.
>
> (Harry)

These teachers began their working lives at a time when EFL chose teachers rather than teachers choosing EFL. None of them set out with a clear idea of the career they would follow. Two took up work outside teaching, Harry drifting through a number of part-time jobs and Paul entering the civil service because he was 'thinking in terms of the limited opportunities that a working-class kid from [City] is supposed to have when he goes to university: become a teacher or join the civil service.'

The others found their way into conventional education, but found the experience unrewarding and dispiriting. Annette 'just felt frustrated and wasn't able to *teach*, because I was concentrating on the discipline side', while Louise found it 'very hit and miss and it's not really *teaching*, in a sense'. Jenny's experience was perhaps the most depressing:

> I used to wake up every morning feeling sick that I had to go in that day...I found it very dispiriting. These people [teachers] were just survivors. Most of them had lost any spark...I hated every minute of it.
>
> (Jenny)

What they all shared in their first exposure to EFL was a sense of discovery in doing something worthwhile, something with tangible outcomes...

> You got a feeling that you *were* actually doing something constructive, that you *were* teaching them something that would be useful for *them*, and have the reward of seeing the progress that they made.
>
> (Annette)

...and a sense of connection with the students:

> I couldn't believe it, it was paradise: small groups of 12 students, highly motivated, polite, interested in the world because they were foreign therefore they wanted to travel. They seemed to have the same kind of interest I had in the world.
>
> (Jenny)

Perhaps, too, there was a sense of being part of a world where there was greater freedom to explore things for themselves in a 'less constrained' occupational context (Jenny).

Coming together

With the exception of Harry, who had previously spent a couple of years teaching English overseas, all the Pen colleagues began their EFL work in the same school, a small, easy-going place run by a 'cerebral' (Harry) philhellene co-director with a love of strong drink and narcoleptic tendencies. Despite the events which were to unfold, he seems to be remembered with mild affection by the Pen staff.

Their view of the director of studies, though, is less positive. Harry's description of him as 'a man's man' is perhaps the least damning representation of his outlook, but the main complaint of the Pen team centres on his indifference to their work. Paul, who joined the school two years after Jenny and shortly before Harry, remembers that his response to everything was 'fine' and things had not changed when Annette and Louise joined the school on temporary contracts five and seven years later, respectively. In fact, research published around this time suggested that this state of affairs may have been fairly common and that there was 'overwhelming evidence that teachers generally receive very little direct assistance and advice from their superiors' (Zeichner et al. 1987: 30).

In situations like this, inexperienced teachers look to their colleagues, and Louise drew strength from her relationship with the group who would later be her colleagues at the Pen:

> It was such a nice atmosphere working with them. I got on with them immediately. I was much more inexperienced then, far, far more inexperienced, and I very much followed coursebooks…They were the sort of people you could talk to and, yes, they were free about talking about what they were doing, and so you could pick up from comments because they'd always been people who'd discussed their work.
>
> (Louise)

In fact, the end of their relationship with the school came shortly after Louise's arrival. They had been aware for some time that things were not quite right financially, and by the end of 1985 they felt that the writing was on the wall. A letter posted to arrive just before Christmas day invited all the permanent staff, Jenny, Paul and Harry, to a meeting at the beginning of January where the director, under pressure from his co-director and the bank, ended their contracts on the spot. Jenny, Paul and Harry told the temporary teachers at the school what had happened at the meeting and, with only one exception, the teachers decided to resign. As a result, all five of the Pen teachers were without work.

Making the break

Finding themselves unemployed, the team acted quickly on foundations that had already been laid. Through attendance at regional teachers' meetings and work as examiners, the core team had already encountered Kate, principal of an internationally known school in a nearby city. She had followed the fluctuating fortunes of their first school with interest and had said, 'Well, if it ever does crumble, phone me first.' When they did, she was as good as her word and within hours all four met in a local pub to agree arrangements for setting up a new school as a sister to the main institution. The new school was to be established on the understanding that it was effectively a junior partner, that Kate would be the nominal principal, and that although it would have almost complete freedom in its day-to-day operational decision-making, key decisions on finance, staffing and so on would be made by the Board of the main school – a condition that would later prove crucial, as we shall see.

Things moved quickly from this point. Kate's generous response meant that Jenny, Paul and Harry received their first salary in April of that year. Even though the budget for buying, decorating and equipping a new school was very tight, Jenny managed to find suitable premises only three or four days after the meeting with Kate and, after persuading the local estate agent that it was worthwhile tracking down the lost keys, was shown around. She remembers that 'it felt right immediately as I walked through the door...The building itself felt friendly, felt good.' While Paul and Harry concerned themselves with timetable issues, Jenny furnished the school in what she describes as 'Mediterranean style'.

The school opened on 1 July and Annette joined the team as soon as work became available six months later, followed a few months on by Louise. This meant that all had a hand in developing its distinctive ethos, but in the meantime a leader had emerged – and the way this happened reveals a lot about the attitudes of the team.

The emergence of a leader

'Effective head teachers', Corrie noted, 'are those who have clear visions for their schools' (1995: 91), and nowadays any prospective principal facing an appointments board had better have their vision statement ready to serve up when the inevitable moment arrives. In the case of the Pen, though, Jenny's emergence was part of a natural process that seemed to arise from recognition that leadership was needed and a shared appreciation of her suitability.

From the start, though, the strength of her vision was clear to the others. As Paul noted, Jenny 'wanted it *right*, and she has a very clear vision of what is right'. And although the three teachers began as a team, role differentiation soon emerged:

> It was fairly sort of collective at the *beginning*, although I think probably from the very beginning Jenny was probably more concerned with the admin side than either myself or Paul. I think the seeds of the present situation were actually there. They just sort of fell in.
>
> (Harry)

Jenny herself sees the emergence of a leader as a necessary evolutionary step if professional momentum was to be maintained but admits that her adoption of the role has left some traces of guilt:

> We found it wasn't efficient to have this ongoing thing because we spent as much time enlightening the next person who was taking over as we ever did *doing* anything. It was consuming so much time that we could have better used that we decided, 'This is *silly* really. One of us has got to take this job on properly and see it through.' It was too bitsy, we weren't achieving anything. And I think this is where bossy boots just said, 'I'll do it.'...I feel a bit guilty that perhaps I did say, 'I want it.'
>
> (Jenny)

In Hargreaves's terms (1995), this represents a clear shift from a rotational maintenance structure to a delegative-rotational structure with the head as *primus inter pares*, characteristic of a collegial culture. The egalitarian-participative political structure remains unchanged and this is an important consideration in the case of the Pen. In fact, seen in this light, Jenny's 'professional efficiency' argument carries considerable weight. Rather than putting colleagues' backs up, it was generally well received:

> I thought Jenny was doing a brilliant job because she had suddenly, almost overnight, from being just a teacher – because none of us were ever more than teachers, we weren't even course director or anything – Jenny suddenly became a manager, part of the management.
>
> (Annette)

Perhaps the main reason for this is that the collegial structure remained undamaged and Jenny's consultative style explicitly values the voices of her colleagues:

> I feel easier if I've got a difficult decision to make if I can go and say, 'Just give me *your* feelings on this. What should I do in this situation?' I value that as well.
>
> (Jenny)

As à Campo (1993: 123) notes, 'Sharing in decision-making gives teachers a greater feeling of ownership which is essential for school improvement' and there is also evidence (Berg 1994: 187) that it reduces the likelihood of educator burnout, an explanation perhaps of the sense of undiminished energy that characterises the school.

Settling down

The prospect of a blank canvas on which to work would appeal to most professionals, but there is a sense in which the palette is already to some extent prepared in their conscious rejection of the way things were done at their previous school:

> What was good about it was we could get rid of all the dross, all the anti-intellectualism that was associated with the staffroom in the other place.
>
> (Paul)

There was inevitably a sense of 'more freedom at the grassroots', a feeling that 'it was all up for grabs, not just the teaching but everything around the school' (Harry), and in the first flush of excitement they decided to give full rein to their professional imagination:

> And so it was a breath of relief that we could come here and do the things *we* believed in...We could let all our ideas free and we could just experiment for a bit. We didn't even have to decide on a style that we went along with, we just had all the freedom in the world.
>
> (Jenny)

As a result, they initially experimented with process syllabuses, popular at that time (Breen 1984), allowing students to set the agenda. However, this was not well received by the students and the team began to feel that things lacked direction. The result, which coincided with Jenny's emergence as leader, was the development of a more structured environment:

> I think out of that chaos you come back to where there's *got* to be a structured view to a certain extent, otherwise the student flounders for one, and they feel that the teacher isn't secure.
>
> (Jenny)

What is interesting about this is not so much the sense of order emerging from potential chaos as a recognition in the team that these changes were at least in part a reflection of their essentially practical and pragmatic orientation:

> We were fairly go-ahead in terms of methodology but we probably all recognised that there's an inherent conservatism as well... We weren't committed, I don't think, to any particular ideology. We were probably all fairly practical.
>
> (Harry)

Once the initial period of experimentation was over, a leader had emerged and the rhythms of school life were settled, the school soon established an excellent reputation with students and inspectors alike. But it would be wrong to underestimate the importance of this first phase. In what follows, I will identify five orientations which underpin their effectiveness as a professional team, the first three of which can be traced directly back to the concerns they had when they first established the school. What will emerge, I hope, is a sense of the dynamic which has sustained members of this team in their professional endeavours throughout their time at the Pen.

Orientation 1   Being person-centred

'Any idea of student-dominated teaching was anathema over there', said Jenny, referring to their previous school, and they set out to change this by making the students the centre of their concerns:

> We wanted it to be very very user friendly. We wanted there to be real and genuine concern for the welfare of the students.
>
> (Paul)

This had implications at all levels. For example, the teachers decided to leave the staff room door open at all times – a very unusual, even radical, decision, as most teachers will recognise:

> It would have been odd for us to close the door on people, because when we set this place up it was supposed to be warm and friendly and actually wanting to go out of our way to help students.
>
> (Paul)

Again, it is interesting to note the extent to which this is a reflection of a more fundamental orientation. All of the Pen teachers mentioned an interest in people as one of the prime motivations for their becoming teachers and Louise's comments are typical:

> What satisfies me about my job is the tremendous opportunity to come into contact with so many different people and to try and establish some sort of relationship with them, some sort of rapport with them.
>
> (Louise)

In order to understand why this matters in terms of their work, it is only necessary to consider what has become the standard view of staffroom interaction:

> The rule of privacy governs peer interactions in a school. It is all right to talk about the news, the weather, sports and sex. It is all right to complain in general about the school and students. However, it is not acceptable to discuss instruction and what happens in class-rooms as colleagues.
>
> (Lieberman and Miller 1990: 160)

Of course the Pen teachers do complain about students – a failure to do so would reflect a lamentable degree of professional indifference – but they also talk a great deal about the students as people, showing concern for their welfare. And there is ample evidence in the data to show that the teachers spend much of their break time discussing students and teaching. In Extract 1, for example, Harry's enquiry about Abdullah evolves into a discussion about course content, student needs and materials, leading to an offer of help for Linda, an inexperienced teacher new to the Pen:

Extract 1

```
H = Harry, L = Linda
001       H:      How was Abdullah this morning.
002               (1.0)
```

```
003    L:    Oh e:r well- I think we're
004          getting somewhere today we're
005          doing (.) you know sort of-
006          'between' 'next' and 'over' that
007          sort of thing and it's
008    H:    Uhuh
009    L:    You know he's (.) he's been oka:y
010          so we're doing sort of starting
011          with the furniture, °describing
012          rooms and that sort of thing. And
013          that's (.) (for beginners. That's
014          why it's good.)
015    H:    Right.
             (180 lines omitted)
016    L:    °(It's a bit xxx)°(2.5) Em (2.0)I
017          also take- (.) Oh (.) and he
018          needs vocabulary building as well
019          and so I've used this (book for)
020          exercises. (1.0) I can also
021          also ⌈(      when I) do this.
022    H:           ⌊It's quite a nice little book
023          though isn't it?
024    L:    Yeah=
025    H:    Harry =Even though it is old.
             (10 lines omitted)
026    H:    There's (.) quite a lot of
027          material around,
028    L:    Yea:h. There's ⌈games and things
029    H:                    ⌊(    ) which I
030          don't mind helping with
```

(T950208–108)

As we shall now see, this interest in the nature of their work is an aspect of the Pen teachers' explicit recognition of the value of continuous professional development.

Orientation 2   Developing as teachers

The Pen team see themselves as teachers first and foremost. Like Annette, they 'love being in the classroom' because that gives them 'the biggest buzz still'. However, their rejection of the anti-intellectualism of their first school arose from a genuine interest in the nature of their work, and early on in the life of their new school they arranged to work together towards an RSA Diploma as a means of furthering their understanding of this:

> I don't know if it made me do any things that I hadn't done before in purely operational terms...but it made me more aware of why I was doing things.

> (Harry)

As Jenny noted, it also gave them 'professional confidence and again it built the team'. This professional confidence is clear in their talk, through their willingness to engage in discussions of their pedagogic practice. Notice how in Extract 2 Paul is discussing an approach to the teaching of tenses involving the adaptation of materials (lines 1–16), how Jenny signals her understanding in line 10 by providing a projected completion of Paul's turn, and how Harry builds his contribution (line 29) on Paul's introduction of possible extensions (line 24):

Extract 2

P = Paul; J = Jenny; H = Harry

```
001    P:      =Yeah. E:M and I think that
002            approach might well work in the
003            future, (.)em (.) for (.) >not
004            necessarily< using this book, but
005            using other texts. And again it's
006            not a question necessarily of
007            inventing your own text, you find
008            a text illustrating past
009            perfect ⌈and then reinvent    ⌉ the=
010    J:              ⌊ And then modify it. ⌋
011    P:      =A part.=
012    J:      =Mm=
013    P:      =And then take it
014              from ⌈ there, ⌉ >I mean< it might=
015    J:            ⌊ Mmm ⌋
016            =be a way in=
017    J:      =Mm
```

```
018   P:   Em (.) all sort of questions come
019        up depending on how (.) perceptive
020        or intelligent people are why do
021        you do it this way. Hehhhh
022   J:   Mm
023   P:   >You know< which is the best way
024        to do it. You could also: bring
025        in terms of how past perfect
026        explains background or just
027        inform- information or ex- or
028        explanation for something. Em
029   H:   Yes that was the basis of my (.)
030        >sort of< (.) analytical
031        thing ⌈I sort of broke ⌉ it down.
032   P:        ⌊Yeah.    Yeah. ⌋
033   H:   E:m (.) and what I did yesterday
034        was that there were basically two
035        kinds of past perfect as far as
036        they were concerned
```
                                      (M941202–570)

This may seem a modest enough exchange, but it represents a stagger-
ing departure from the sort of staffroom talk that has featured in
research for over a quarter of a century (e.g., Hammersley 1984;
Hargreaves 1981; Kainan 1994). In Supovitz's study of team-based
schooling, for example, talk about teaching was the one expected outcome that failed to materialise, leaving the researcher to claim that
'groups need to develop a culture of institutional practice, one that
encourages them to continuously and safely identify, explore and
assess instructional strategies that show promise of success with their
students' (Supovitz 2002: 1618).

Orientation 3   Staying fresh

Shared exploration is typical of professional exchanges in the Pen
and reflects the team's determination to stay fresh. The sense of stagnation in their previous school, where 'nothing ever changed'(Jenny)
was something they were determined to avoid in their new school, so

they made sure that they took it in turns to take responsibility for professional development:

> It's vital to us staying fresh.... If you've got somebody whose focus *is* our professional development, who's sort of keeping us on our toes... It's terribly important to *us* as professionals, otherwise we *do* feel that we get into the daily grind of the full five hours a day every day.

> (Jenny)

The evidence of this commitment, however, is more deeply embedded than even this might suggest, and a close analysis of their talk reveals a metaphorical orientation which permeates their talk. Positive aspects of professional practice are expressed in dynamic metaphors involving action and change, while negative features are represented in static terms, often involving accumulation and stagnation (for further evidence of the importance of metaphor in teacher talk, see Mann, this volume). The following extracts, taken from different days and times, use a single image, the box or container, to illustrate this.

In Extract 3 Paul argues against a new school system on the basis that it will involve simply piling up pieces of paper instead of exploiting materials and exchanging ideas (lines 1–11), concluding his point by representing the result as a 'sump' (line 13):

Extract 3

P = Paul; H = Harry

001	P:	It's just going to be a <u>heap</u> of
002		things. So I- <u>I</u> thought it- it
003		m<u>i</u>ght be an idea just to-
004		(0.5)
005	P:	just to look at d<u>i</u>fferent ways of
006		exploiting <u>a</u> piece of material
007		which
008		(1.0)
009	P:	which we may <u>know</u> about, and sort
010		of pooling id<u>ea</u>s rather than
011		pooling pieces of paper.
		(19 lines omitted)
012	P:	Yeah. And I think that is a sort
012		of a danger >you just< end up with

```
013                   a sump=
014        H:         =Yeah=
015        P:         =of material
```

(M941111–65)

Eventually, the others adopt the same metaphor and his argument is carried. Unsurprisingly perhaps, given that the idea of simple accumulation has negative connotations for this group. In Extract 4 Paul's complaint about teaching lists of stock phrases receives a sympathetic hearing from Harry, who represents these as 'boxes' of expressions (line 12):

Extract 4

```
P = Paul, H = Harry, A = Annette
001        P:         I hate that. (0.5) Stock phrases
002                   (.) nonsense.
                      (23 lines omitted)
003        H:         °Yeah.° Quite clearly there are
004                   some coursebooks that encourage
005                   that approach.
006        P:         Absolutely.
007        A:         Mmm=
008        P:         =Well look at proficiency how it's
009                   still t-taught. And first
010                   certificate.=
011        H:         =And boxes of useful: (.)
012                   expressions.
```

(T950301–513)

Orientation 4 Seeing the joke

Staying fresh also involves maintaining a positive perspective, and humour has an important contribution to make to this. Humour is woven into the very fabric of talk in the Pen staffroom, and although there is no space here to explore its many dimensions (for a fuller treatment of humour, see Richards 2006: ch. 4), the team's views are significant. Harry says that fitting in 'might be something to do with how seriously you take yourself' and Jenny is even more explicit:

> The ones that make it here are the ones that don't take themselves too seriously.

(Jenny)

They even recognise one of the commonest forms of humour in the staffroom, playing off this for humorous effect:

Extract 5

H = Harry; A = Annette; K = Keith; P = Paul

001	H:	The amount of deliberate
002		misunderstanding that goes on in
003		this staffroom=
004	A:	=Heheh=
005	K:	=HEHEH HEHEH!
006	H:	Heheh
007	P:	What do you mean by that.

(T950301–662)

Orientation 5   Being together different

It's clear that the Pen team have a lot in common – that's why they set out on this venture together – but professional effectiveness is blunted by bland uniformity, as Hargreaves has recognised:

> Collaborative cultures require broad agreement on educational values, but they also tolerate disagreement, and to some extent actively encourage it within those limits.

(1992: 126)

The final element I should like to highlight as essential to the success of this team is a recognition of the importance of difference within a creative context. The words of Jenny and Louise capture the essence of Hargreaves's observation with uncanny precision:

> It *might* be something to do with the slight difference in our personalities, I think. That there's *enough* difference for conflict of a certain kind all the time. I think that's quite good, that we can strike ideas off each other and don't just completely blandly agree.

(Jenny)

> That's what I say, that's what's so good about working here, the fact that we do get on so well even though we do have different ideas. We respect each other's ideas.

(Louise)

The dimensions of difference and similarity are too complex to admit of easy representation, but some sense of their relationship can perhaps be captured by highlighting the relationship between understanding and difference that informs the working partnership of Paul and Harry. Both are unequivocal about the almost instinctive understanding that exists between them. Here is Paul on the subject:

> I think Harry's quite important to me because I think we enjoy talking to each other. I think we do, so it's quite complex.... We understand each other completely here, Harry and I.
>
> (Paul)

One might expect, then, that their approaches to teaching are similar, but nothing could be further from the truth. While Paul is very much the 'actor' in class...

> The thing about me is that it's the performance I'm interested in. It really is performance and anything related to it.
>
> (Paul)

...such an approach is anathema to Harry:

> I'm fairly *casual* in class and very cards on the table...People often *do* and I have done – probably everybody has done – allowed the methodology, the performance aspect, to take over from – because you forget the idea of actually what you're supposed to be doing, you're teaching them how to learn the language.
>
> (Harry)

Although the link between teacher beliefs and practice should not be underestimated (see Garton, this volume), the evidence of the Pen suggests that common fundamental values are more important than teaching style for effective professional collaboration. Differences are, in fact, to be welcomed, but only if there is a solid foundation of mutual respect.

The Pen legacy

> The big thing is, is somebody going to rationalise us?
>
> (Paul)

I have tried in this chapter to tell the story of a school and to capture something of its character. Like all stories, this one has an ending – an answer to Paul's question.

The Pen was a successful school on all counts, but it was the younger and smaller relation of a school in a nearby city where, ultimately, all the decisions were made. The links, unfortunately, were not strong: the Pen team had 'never had much contact' with the other school (Harry, Paul) and 'there was never any feeling of being their colleagues at all' (Louise). So when part of the larger school began to lose money and sacrifices were needed, the Pen was in the firing line. It was felt that money could be saved by diverting their work to the main school, which in turn would bolster student numbers in the part which needed revitalising, so the Pen was closed.

At least that's the story as I heard it. By then I had left the Pen and I learnt of the school's fate from Harry. I still regret the passing of such a happy and successful enterprise and the breaking up of a team of good people, but I'm also grateful that I was able to spend time with them. For that reason I describe what follows as their legacy.

In terms of what we know about schools and their organisation, the Pen seems to offer an unusual example of a genuinely collaborative team, working with their students' interests at heart and engaged in a process of continuous exploration and development. To understand how important this is, we need to appreciate the tension that arises for many teachers from the conflict between their natural disposition and the system of which they are a part. Hargreaves (2001: 1069) captures this well:

> In many ways, schoolteaching has become an occupation with a feminine caring ethic that is trapped within a rationalized and bureaucratic structure.

Team-based schooling has been tried, and the results have been largely positive. Supovitz's study (2002), for example, revealed that teams produced a more student-focused school culture, a shift of decision-making to teachers and an increase in teacher accountability, all features of the Pen school. However, as I noted earlier, an expected focus on curricular and instructional issues was not found, and in this respect the Pen is different. It is not possible to say definitively why this might be, but it may have something to do with a corollary of the freedom to act: the freedom to fail. Where such failure results in the loss of students and thereby of income, the stakes for the teachers involved are high and the incentives to improve professional practice are correspondingly great. But perhaps I do the Pen teachers a disservice: their commitment to professional development was there from the start and their engagement in it seems to have been a source of satisfaction to them.

There is, of course, no recipe for a successful school, but (at the risk of excessive simplification) I should like to offer the following brief summary of some of the important lessons that might be learnt from the work of the Pen teachers:

- It is possible for like-minded teachers to establish a new school and work together in harmony to make it successful.
- Democratic decision-making and collaborative effort can coexist with the presence of a leader, but ideally such leadership will emerge naturally.
- A commitment to continuous professional development is reflected in daily professional exchanges aimed at sharing good practice.
- Serious professional concern and a sense of humour (including the ability to laugh at oneself) are not mutually exclusive.
- Successful collaboration depends on a combination of shared values, individual difference and mutual respect.

These are not surprising conclusions, but they may be important ones, not least because they are true of at least one school at one time. They are not, therefore, counsels of perfection but a reflection of something that actually existed and might therefore exist again. Sometimes it is legacy enough to show that a thing was possible.

References

à Campo, C. 1993. 'Collaborative school cultures: How principals make a difference'. *School Organisation*, 13(2): 119–27.

Berg, B. D. 1994. 'Educator burnout revisited: Voices from the staff room'. *The Clearing House*, 67(4): 185–8.

Breen, M. P. 1984. 'Process syllabus for the language classroom'. In C. J. Brumfitt (ed.), *General English Syllabus Design* (ELT Documents 118). London: The British Council.

Corrie, L. 1995. 'The structure and culture of staff collaboration: Managing meaning and opening doors'. *Educational Review*, 47(1): 89–99.

Hammersley, M. 1984. 'Staffroom news'. In A. Hargreaves and P. Woods (eds), *Classrooms and Staffrooms: The Sociology of Teachers and Teaching*. Milton Keynes: Open University Press, pp. 203–14.

Hargreaves, A. 1981. 'Contrastive rhetoric and extremist talk: Teachers, hegemony and the educationist context'. In L. Barton and S. Walker (eds), *Schools Teachers and Teaching*, Lewes: Falmer Press, pp. 303–9.

Hargreaves, A. 1992. 'Cultures of teaching: A focus for change'. In A. Hargreaves and M. G. Fullan (eds), *Understanding Teacher Development*. London: Cassell, pp. 216–36.

Hargreaves, A. 2001. 'Emotional geographies of teaching'. *Teachers College Record*, 103(6): 1056–80.

Hargreaves, D. H. 1995. 'School culture, school effectiveness and school improvement'. *School Effectiveness and School Improvement*, 6(1): 23–46.

Huberman, M. 1992. 'Teacher development and instructional mastery'. In A. Hargreaves and M. G. Fullan (eds), *Understanding Teacher Development*. London: Cassell, pp. 122–42.

Kainan, A. 1994. 'Staffroom grumblings as expressed teachers' vocation'. *Teaching and Teacher Education*, 10(3): 281–90.

Lieberman, A. and Miller, L. 1990. 'The social realities of teaching'. In A. Lieberman (ed.), *Schools as Collaborative Cultures: Creating the Future Now*. London: Falmer Press, pp. 153–63.

Richards, K. 2006. *Language and Professional Identity: Aspects of Collaborative Interaction*. Basingstoke: Palgrave Macmillan.

Stenhouse, L. 1980. 'The study of samples and the study of cases'. *British Educational Research Journal*, 6(1): 1–6.

Supovitz, J. A. 2002. 'Developing communities of instructional practice'. *Teachers College Record*, 104(8): 1591–626.

Zeichner, M., Zeichner, B., Tabachnik, R. and Densmore, K. 1987. 'Individual, institutional, and cultural influences on the development of teachers' craft knowledge'. In J. Calderhead, (ed.), *Exploring Teachers' Thinking*. London: Cassell, pp. 2–59.

Reflections on New Horizons

Jerry Talandis Jr.

As an English teacher firmly ensconced in the 'New Horizons' stage, I can confirm with confidence that experimentation and reassessment are indeed major themes I'm currently facing in my career. The chapters by Quirke, Mann, and Richards have helped me reflect on these in productive and insightful ways. In the commentary that follows, I offer up my own story in hopes of highlighting how the research done by these authors has resonated in the life of one teacher. I'll begin with some background information, looking briefly at my professional development arc. Next, I will delineate some issues at the forefront of my mind these days, then provide commentary on each chapter in light of them. Finally, I'll offer up some conclusions on insights gained and lessons learned.

I've been teaching English in Japan for the past 14 years. I began as an assistant language teacher in the JET Programme, a governmental organization designed to import young native English speakers to help teach English in the secondary school system. After three years of 'team teaching' junior and senior high school students, I landed a job at the Toyama College of Foreign Languages (TCFL), a small two-year vocational school specialising in English study and have been here ever since. Our school primarily serves students (mostly recent high school graduates) helping them prepare for higher education or for careers where knowledge of English is seen as an advantage, in fields such as travel, hotel management and education. In my more than ten years here, I have taught the entire spectrum of our curriculum, courses ranging from the basic skills, TOEIC and STEP test preparation, and content-based courses such as world affairs, drama and video production.

Although I've been teaching for a long time, formal teacher training is not part of my educational background. At university in the United

States I majored in psychology and graduated with a bachelor of arts degree in 1986. Following that I worked for six years as a counsellor in a mental health programme, helping adults with severe mental illness make the transition from institutionalised to community-based care systems. While I greatly enjoyed working with clients and helping make lives better in small ways, the overall stress of the profession eventually prompted me to change careers. My interest in travel and adventure at the time led me to the JET Programme, which I viewed as an apprenticeship of sorts, a vehicle allowing me to develop the skills and experience I would need for a new career. My new career choice became firmly set when I got married to a Japanese national, settled down and began raising a family in Japan. Financially and professionally speaking, there was no other viable choice but to continue down the path of becoming a full-fledged English teacher, a career I had never imagined doing when I was a child.

For my first eight years or so, it was really a case of 'learn by doing'. I received some limited teacher training from various workshops, but for the most part it was simply learning from trial and error in the midst of work. After a while, it became apparent that the only way forward, both personally and professionally, was to enhance my credentials and look for a better, higher-paying job, either at a university or as the owner of my own school. In either case, a masters degree was essential, and given my work and family responsibilities, the only choice was to attain this credential via distance learning. In the spring of 2001 I enrolled in the Aston University MSc in TESOL programme, and after five years managed to attain my degree.

With a degree in hand and years of teaching experience behind me, the question foremost in my mind these days is 'what now?' With the recent birth of our second child, I am feeling more pressure than ever to provide a better living for my number one priority, my family. Professionally, I'm at a crossroads. While I love my current job, I can't stay indefinitely given the limits of what it can offer me financially and professionally. Unfortunately, due to the rural nature of the area I live in, employment opportunities, especially at the university level, are few and far between. I feel stuck in a sense, not knowing how I'll be able to sustain the deep roots we've planted in our community. Something will have to change: either I uproot my family and move to another area, or I take the plunge into self-employment. Either road holds uncertainty, and as a result, anxiety over my future career looms darkly, casting a shadow over my life. The primary issue, therefore, is resolving this conundrum I've put myself in. What is the best way forward?

With this backdrop I read the chapters by Richards, Mann, and Quirke, searching for some insight into how to resolve my situation. What I found confirmed my instincts, that the best way forward was to make personal and professional growth a top priority. Themes of introspection, articulation, communication and collaboration present in each chapter spoke to this growing realisation in practical and insightful ways.

At a practical level, Richards's account of establishing a new school addressed my question of which career track I should take. I've gone back and forth on the idea of beginning my own school, to shifting my professional life 'onto a precarious balance of risk and reward'. The complete account of a particular case study was most helpful in providing me with new information to reassess this particular career option. The way in which the data was presented allowed me to experience the Pen school teachers' situation vicariously and thus learn a great deal from it, despite the specific differences in our respective circumstances. For example, most profound for me was the section where all of the teachers simultaneously lost their jobs. That must have been an incredibly stressful time, one I could relate to, having been through something similar. Seeing how they all managed to come together and work through it provided me with inspiration I can use to resolve my own career conundrum. Key here was the emphasis on collaboration and communication. Prior to losing their jobs, the five teachers had all gotten on well despite less than ideal working conditions. They supported each other and made efforts to network with others in their profession. This collegial support turned out to be their saving grace when their school went under. The change they went through, however stressful it may have been, turned out to be a blessing in disguise. This is a story that will never grow old with me, and I was heartened to see it emerge from within a work of academic research.

The chapter by Mann on metaphor usage in cooperative development (CD) recalled to mind my own personal experiences with this form of teacher talk and helped me look at them in a new way. As part of my masters studies with Aston, I took a course in this approach to professional development, but with a twist: instead of conducting sessions face to face, I engaged in CD via synchronous instant messaging with two colleagues living in areas far from my own. The sessions I conducted proved to me that genuine self-development could take place over the Internet, without the need for face to face communication. No doubt a degree of subtlety was lost in this mode, in addition to the overall slowness and occasional technical problems. However, after a period of adjustment, the advantages of instant transcripts, access to computer resources, and added reflection time (via a simultaneous 'real-time journal' and follow-up emails) become apparent. CD is thus a method open to all practising teachers with an Internet connection.

The role of metaphor was not something I ever considered when doing CD, but now looking back at some of my old data transcripts, I can see the role it played. In the following extract, taken from an email discussion reflecting back on a CD session I completed as *speaker*, notions of 'energy' and 'family' provided a powerful means of expressing deeply felt emotion when considering my future career prospects:

> So, as you can see, I had a really good session. It was surprisingly effective. I didn't know what I'd talk about until the moment I started off about my future. Like I said above, that decision was totally based on my intuition. As so often happens, when I follow it, I end up in good places.

> When you got me to focus on why I love working at TCFL, that really had a liberating effect on me. I felt a rush of positive energy that had the effect of canceling out some of the anxiety I was feeling about this topic. Then the source of my anxiety became clear – I really dreaded leaving this 'family' atmosphere. I hadn't thought of it in those terms before, so realizing that was helpful.

> I think that connecting with my fears had a positive overall effect. I know that everything will work out, and affirming that felt great.

'A rush of positive energy' is a phrase I often use unconsciously to illustrate what the movement of strong thoughts and emotions feels like for me. The quotation marks surrounding the word *family* indicate I was using this term consciously to describe deep affinity with my co-workers. As I can now see, these metaphors played a role in my realisation that moving on in my career would involve a great deal of personal change.

Quirke's chapter on supporting teacher development on the web resonated with me through its emphasis on overcoming professional isolation, in my experience a common problem facing teachers interested in moving forward in their careers. His model of teacher knowledge acquisition not only helped me conceptualise how I learn in constructivist terms, but confirmed for me the importance of active involvement within a community of practice. The nine keys of practical advice offer direct, simple and commonsense ways of establishing and maintaining a network of collaborative relationships. They serve as a road map for utilising positive energy to enhance communication with colleagues. With the insight and tools this chapter provided, I realised that professional isolation is a now a choice I make, and not something over which I have little or no control.

To conclude, the chapters by Richards, Mann, and Quirke all expounded on themes that have spoken to me deeply as I reassess my future career plans. By highlighting themes of introspection, articulation, communication and collaboration, I felt encouraged to look within, find my talent and passion, then get actively involved in sharing it with others in my profession. I don't know how things will eventually play out, but I'm convinced that the way forward unfolds naturally as thoughts are clarified and expressed, as connections are made, as honest service to others is rendered. It's comforting to know that as long as I maintain my focus on doing these things, opportunities for professional growth will arise, and my future will unfold as it should. In the end, the research presented in this section has left me feeling empowered and excited for what lies ahead.

Part IV

Passing on the Knowledge

Paul

Paul's had enough. Like most of his generation, he thought the Direct Method was a genuine step forward, felt that *Strategies* was a breath of fresh air and embraced Communicative Language Teaching with a passion, but now he doesn't know what to think. It's a post-method world nowadays and whenever he draws on the old certainties to make a point in a staff meeting or pass on a word of advice to a young colleague, he can almost hear the sighs of resignation.

He can cope with that – after all, they missed out on the Golden Age and have to compensate somehow – but it's harder to come to terms with the sense of ingratitude he feels about the rejection of all his efforts over the years. He was *never* a colonialist; all he wanted to do was help people in their struggle to master a world language. Now it seems he was party to a hegemonic plot designed to disempower the very people he thought he was helping, and what's worse (this really hurts) it turns out that English isn't even his own language any more. What are *Englishes* for heaven's sake?

Fortunately, his students are still wonderful, so that'll be a comfort over his last five years in the job. And when he retires he can spend all his time working on the garden, which is where his heart is now. In fact, he spends all his non-classroom time in school thinking about dahlias, reading garden magazines and sketching new garden layouts. Colleagues bring their gardening problems to him on a regular basis and he's become something of a horticultural guru. Perhaps he chose the wrong career in the first place.

The final phase

For obvious reasons, Paul's not reading this book (it has little to say about dahlias), but he is apparently typical of many teachers who have remained in the classroom. That at least is the conclusion of the research that we have to date. In reporting on the VITAE project, a government-funded four-year collaborative study involving working with teachers to identify 'factors that may affect their work and lives over time and how these factors may, in turn, impact on their teaching and subsequent pupil progress and outcomes', Day and colleagues describe the final phase in terms not very different from those of Huberman:

> The final phase (increased concern with pupil learning and increasing pursuit of outside interests; disenchantment; contraction of professional activity and interest, disengagement, serenity).
>
> (Day, Stobart et al. 2006: 174)

They also refer to Farber's finding (1991) that career development is often accompanied by a progressive sense of inconsequentiality.

It's not a happy picture, and even though the situation in TESOL may be different in some respects, many teachers in the profession will recognise Paul, or at least elements of him, in some older colleagues. The issue here, however, is not whether teachers in the final stage of their career manifest characteristics of the relevant life phase – it is inevitable that they should – but whether there is anything we might usefully say about the part that experience plays in the development of our profession. To do this we need to think beyond the obvious.

Traditionally, the more experience you have, the more valuable you become to those in your group as a repository of knowledge and wisdom; you become a resource to be treasured and are respected accordingly. However, in society or a profession where knowledge is accessible to all and where approaches, methods and procedures change with bewildering rapidity, experience does not serve the same purpose and the inevitable slowing down that comes with age makes it increasingly difficult to keep up. Ironically, the form of knowledge that is of most value in terms of performance cannot be passed on and actually serves to inhibit receptivity to change. Leinhardt characterises the expertise associated with *situated knowledge*, accumulated through extensive experience, as a bank of detailed information (about students, materials, curriculum, etc.) and a large repertoire of behaviours. Resistance to change on the part of teachers who have developed a rich stock of such knowledge, she argues, should perhaps not be perceived as 'a form of stubborn ignorance or authoritarian rigidity but as a response to the consistency of the total situation and a desire to continue to employ expert-like solutions' (Leinhardt 1988: 146).

The issue, then, is neither one of accumulating knowledge and then passing it, nor one of criticising older teachers for resisting change; it is rather a matter of exploring how knowledge can be shared and experience pooled in productive and creative ways to the benefit of all involved. This is what Passing on the Knowledge involves and, in this respect at least, research into teachers' lives is more encouraging:

> Teachers overwhelmingly valued supportive colleagues and the sharing of good practice when asked to discuss what was important within the teacher and school culture. In addition, of the influences most likely to impact positively on their work, teachers said that supportive colleagues or a sense of community in the school was key.
>
> (Day, Stobart et al. 2006: 183)

In this final part, then, we break with the neat trajectory that the book has so far followed. There will be no evidence of disengagement, serene, bitter or otherwise; instead, the writers address the challenges facing those who wish to share their experiences and understandings with others. These chapters mirror those in the opening section. Both are concerned with the exchange of knowledge, but here we are less concerned with the receipt of knowledge than in the ways in which it might be passed on in a process of mutually supportive exchange.

Three perspectives

The three authors in this section have all, in their different ways, wrestled with the challenges of finding discourses that would enable them to promote the sharing of knowledge and understanding. Kuchah Kuchah, a relatively young professional in a position usually occupied by more senior colleagues, sets his story in the context of a postcolonial educational culture in which embedded asymmetries reinforce inherent conservatism and legitimise the imposition of inappropriate practices. Creative responses to this, he shows, depend on establishing modes of communication that circumvent the limitations arising from power differentials and allow the development of productive relationships with local practitioners and practices.

Kuchah's early experiences in teaching older students exposed him to tensions arising from the need to reconcile the exercise of his authority as a teacher with the obligation to conform to local customs, especially those relating to the respect due to age; but more importantly they revealed a failure on the part of the system to recognise and respond to the sociocultural diversity of his country. It was the realisation of this that underpinned his radical decision to develop community-based content, which in turn resolved the tensions he had initially faced.

Having been appointed one of his country's youngest inspectors, he was faced with the same problem in a new guise: how to change Ministry practices without overtly challenging his elders and how to establish a productively symmetrical dialogue with teachers who now saw him in an essentially judgmental role with power over their futures. Only by exploiting the potential of discourse communities as platforms for exchange was he able to respond to these challenges, and he writes now from the perspective of someone who, through subsequent course-book writing and postgraduate studies, has extended his understanding of how this work might be pursued further to meet the needs of teachers and students in his country.

Sue Wharton's focus is narrower but its reach potentially even broader. Many teachers are used to exchanging their knowledge and understanding in the context of staff meetings, local teachers' groups and perhaps national or international conferences, but few try their hand at writing, which is perhaps surprising given the nature of their calling. As Wharton notes, this is widely recognised, but there seems to be relatively little advice available to teachers who would like to extend their sharing in this way.

Wharton offers practical advice to such teachers, ranging from a discussion of the outlets that are available to an exploration of the challenges facing the aspiring writer. Drawing on teachers' own voices from her work with prospective authors, she addresses issues of authority and insecurity, the reader–writer relationship and the delicate judgement of how much to tell. She also uses insights into her own experiences to explore the more subtle psychological adjustments that are associated with the assumption of a writer's identity.

There are no simple prescriptions in Wharton's chapter – the last thing that language teachers need is a recipe for effective writing. Her paper is designed rather as an illustration of what is possible and how it might be achieved, exploring the challenges – linguistic, professional and psychological – that seem at the moment to be inhibiting teachers who might otherwise have valuable insights to share. If her contribution succeeds in persuading just a handful that publishing is within their reach, it will have been instrumental in enriching the profession.

The discoursal heart of Wharton's chapter is Hoey's Problem-Solution model, which is also where Edge's story starts. This final chapter in the collection looks back on a lifetime in which the author has worked as a teacher and with teachers in exploring the potential of personal and professional development. As the chapter demonstrates, the journey has been a complex one and the exploration continues. Even the basic model has been refined over time to better reflect teachers' circumstances and the ways in which they can better understand them.

One of the most interesting aspects of Edge's chapter is his exploration of how the three core strands of his work interweave to represent the intellectual fabric of his professional life. For him the Problem-Solution pattern is more than a way of understanding discourse, it is a way of better understanding the nature and potential of action research, and the concerns of the latter with continuing professional growth are realised through the distinctive discourse practices in Cooperative Development. What makes Edge's contribution to this collection distinctive and special is the way in which it realises the professional life through its discoursal complexities without shirking the conceptual, practical and moral challenges that this entails or retreating into the pretence that these can easily be reconciled.

Will Edge's interest in the relationship between Situation and Evaluation win through and become a priority? He poses the question half way through his chapter, but does not provide an answer. The response is to be found in the unresolved synthesis at the close, in the universal search for coherence, and in the author's determination to pursue goals worthy of our profession, however elusive they may be.

In her comment on the value of the Problem-Solution model in teacher education, Fotini Kuloheri makes a useful connection between Edge's chapter and the opening part of this collection, and her suggestion that AR reflects a natural orientation in teaching is an interesting one. She also offers a teacher-writer's perspective on Wharton's chapter and makes a convincing plea for the study of writing for publication in Masters programmes. Her most poignant comments, however, are to be found in her response to Kuchah's chapter, where she reveals challenges in the Greek educational system that parallel those in Cameroon. What emerges from the experiences of these two writers is a strong sense of the need for committed individuals to work for change, whatever the obstacles.

Conclusion

'Share your knowledge', says the Dalai Lama, 'It's a way to achieve immortality.' A simple enough prescription, but as these final chapters show, first you have to find a common language. None of the three authors here has found the process of sharing their knowledge straight-forward and for each of them the struggle has taken a different form. There is perhaps in Wharton's chapter a sense that the issues have largely been resolved and that the ways forward have become clear, but for Kuchah and Edge the conundrums remain; and in this there is perhaps an important lesson that can be understood only from the perspective of the final phase in a teacher's career trajectory. Those who have left teaching to take up senior positions in management or administration will be able to reflect on the extent to which their professional targets have been met, but the achievements of the classroom teacher lie in the hearts, minds and actions of others.

References

Leinhardt, G. 1988. 'Situated knowledge and expertise in teaching'. In J. Calderhead (ed.), *Teachers' Professional Learning*. Lewes: Falmer, pp. 146–68.

Day, C., Stobart, G., Sammons, P. and Kington, A. 2006. 'Variations in the work and lives of teachers: Relative and relational effectiveness'. *Teachers and Teaching: Theory and Practice*, 12(2): 169–92.

Farber, B. 1991. *Crisis in Education*. San Francisco, CA: Jossey-Bass.

10
Developing as a Professional in Cameroon: Challenges and Visions

Kuchah Kuchah

Introduction

This chapter describes a professional experience which, albeit personal, is representative of the realities of professional development in Africa. My purpose is to demonstrate from a practical stance that developing within the teaching profession in sub-Saharan Africa is fraught with challenges and realities that shape and are shaped by a teaching/learning culture of power relationships which constantly places the evolving teacher at the crossroads of discourse constraints. To achieve this purpose, I begin with a very brief overview of the transition from a colonial system of education to a modern system developed against a background of colonialism, setting the scene for my exploration of how the discourse patterns of a developing professional are constantly in a flux. Although my narrative is based on my personal experiences working in Cameroon, I am confident that it will raise questions and issues that relate to other African countries, even allowing for the considerable cultural differences among them.[1]

The historical relationship between Cameroon and two former colonial powers (France and Britain) places it at the crossroads of linguistic/curriculum (con)fusion. After independence in 1961, La Republique du Cameroun (French Cameroon) and Southern Cameroons (British Protectorate) decided to come together as one country. This meant that both parts were each to bring along a system of education and language inherited from their colonial masters. For a very long time, the two subsystems of education in Cameroon simply followed a curriculum emanating from the colonial period. It was not until 1995 that a national forum on education was convened with the aim of redefining educational policies that will reflect the needs of the Cameroonian society (Law on Education 1998).

However different the learning objectives of both subsystems might have been at independence, one common denominator was the teaching/learning culture that arose from the 'ashes' of colonisation. The generation of teachers who took over from the colonial teachers, and who became mentors to the generations after them, seemed to have inherited, from their predecessors, a belief that 'the ears of the African learner are on his back', so that a little flogging will play the trick! With a rising concern for the rights of children hurried down their throats, they 'softened' down to bullying and then to nagging! As a child, I so wanted to become a teacher so I could return the flogging of my primary school teachers to other learners, but as I grew up and met different teachers, I came to realise that the teacher could have a relationship with learners that is not entirely top-down.

Received wisdom

My professional experience starts in 1996 when, graduating as a teacher trainer, I was sent to teach English and English teaching methodology in the far north province of Cameroon. As a student-teacher in the Higher Teacher Training College, I had been stuffed with a considerable dose of child and adolescent psychology, of pedagogic principles and methodology; I had a good knowledge of the audio-lingual and CLT methods, of task-based methods and a repertoire of strategies for implementing these methods. But, as seems to be the case with most young people getting into the teaching profession in a context that is basically examination oriented, these methods and theories had value to me only in as much as their reproduction would help me make good grades and graduate. I could list and explain Piaget's stages in cognitive development, but these had nothing to do with a genuine awareness of their implications for the classroom.

As a young teacher trainer for primary school teachers, my training claimed to prepare me for the realities of teaching young learners so that I would be in a good position to train future teachers for that level. But such training, in addition to the shortcomings already mentioned, had the further weakness of not preparing me to be able to cope with adult learners who were my direct learners. Teaching in a training college of more than 300 trainees, the youngest of whom was four years older than me was the first challenge that impacted on my discourse inside and outside of the classroom.

In a society where age is an important determining factor of power relationships, where the young are seen as immature, naïve and ignorant while the elderly are seen as mature, reasonable and wise, it becomes incumbent on teachers to impose their personality in such a way as not to destroy the cultural status quo. The verbal and non-verbal interactions of teachers tend to be influenced by the need to present self-image without jeopardising the social fabric of the community (Diamond 1996). Teaching in Cameroon inadvertently demands a certain amount of multicultural awareness because with over 250 tribes, cultures and local languages, what may be seen as culturally appropriate in one part of the country will be inappropriate in another part. For example, coming from the north-west province of Cameroon, I had learned never to stretch out my hand to an older person unless he offered to shake hands with me first. In the far north province, I had to cope with even children offering to shake hands with me. This example not only illustrates the opposing realities that underlie social and consequently, classroom interaction in Cameroon, but also provides a basis for understanding my initial discourse constraints as a novice teacher trainer.

Having worked as a teacher trainer to adult trainees under the demanding cultural conditions placed upon me, challenges that were not part of my training, it seems to me now that initial teacher training in Cameroon pays little attention to the sociocultural diversity of the country and that training content is tailored to meet Western pedagogic innovations which can sometimes be at odds with the demands of our specific contexts. Not having been trained to deal with adult learners in a culturally demanding context, I was caught between a desire to be an accepted member of the community and the need to assert my authority over my trainees. The latter desire will probably seem strange to some teachers, but for practitioners in Africa, where there is a complex web of power relationships, it is quite in order. The discourse of young professionals within this kind of context is more likely to aim at reinforcing the status quo at the expense of making connections with the realities of teaching; consequently, the discourse of pedagogy is sacrificed for the discourse of power which inevitably builds an unhealthy asymmetrical relationship not only between teachers of different generations but, most especially, between teachers and learners (for a more detailed discussion of the discourse of power, see Fairclough 1989: ch. 3).

Many factors account for this kind of asymmetrical relationship between teacher and learner. One of these is the fact that over the years, the educational tradition I have described above has positioned the teacher as the sole provider of knowledge, and although initial training may insist on an interactive approach to pedagogy, novice teachers still tend to go back to traditional approaches. In addition, because of the ailing economies of African nations, education presents itself as being the most reliable means to a livelihood, parents are very keen and some-times impatient about the education of their children. For this reason, the teacher is seen as a master and although teachers may be poorly paid, they command a lot of respect, especially in rural communities. The result is that learners are made to depend on the teacher for the knowledge they acquire and although other sources may be open to them, the teacher's verdict is considered to be final.

Another reason for this power relationship emanates from the under-resourced nature of our schools. In state schools, it is common to find situations where less than 10 per cent of students can afford the prescribed textbooks. The teacher therefore becomes the sole mediator between textbook, curriculum demands and students. With the added burden of teaching classes of over 140 students, teachers tend to apply traditional methods, which are most often accompanied by a subtle dictatorship over learners. Relying more on my intuition rather than on my training, I sought to develop a teaching/learning relationship based on hands-on experiences and therefore shaped my discourse to meet the immediate needs and convictions of my learners.

Connecting with local practice

I have already mentioned the national forum on education which marked the turning point for education in Cameroon. In order to better explain my role as teacher trainer for English in a French-medium teacher train-ing school, it will be necessary to briefly explain the impact of the Forum on the teaching and learning of the two official languages of Cameroon. Amongst other things, the 1995 Forum laid the groundwork for the implementation of official bilingualism, that is, English and French at all levels of French- and English-medium schools, respectively. Because both languages are the official languages of Cameroon, with each having equal status (Constitution 1996) they are usually referred to not as sec-ond or foreign languages, but as either 'first official language' or 'second official language' depending on which of them is the main tool of com-munication for a given individual. One recommendation of the Forum was to implement the teaching/learning of English in French-medium schools and French in English-medium schools at the nursery and primary levels, and this meant that teacher training colleges were to prepare student teachers to be able to teach both languages in their classrooms. It is also worth noting that although bilingualism started as a political option aimed at promoting national unity and integration between French- and English-speaking Cameroonians, it has become a vogue goal for many educated Cameroonians. There is an observable upbeat attitude in parents and children towards the two languages with English taking the lead because of the imposing socio-economic, technological and political transformation of the world by the United States and United Kingdom.

My mission, therefore, as a teacher trainer, was to instruct my trainees in different ways of teaching English language to children in French-medium primary and nursery schools. But how could they be trained to teach a language they were not sufficiently confident with? I was therefore compelled to pay more attention to teaching them the English language itself rather than methodology. This meant that I had to sacrifice an important part of my job description for another which was, for my particular context, more relevant. And as my trainees were generally at the lower elementary level, I needed to pace input as one would normally do with primary school ESL learners. For an examination-centred syllabus, this can be very frustrating for a teacher who is caught between administrative sanction and pedagogic reality. But by its relevance to the immediate realities of the context within which both teacher and students operate, it solves not only the immediate linguistic needs of the learners, but also creates a healthy psychological relationship between both parties which is invaluable for a pedagogy of engagement. In this way, the relationship between linguistic practices and psychological processes was mediated by both cultural and interpersonal categories and this mediation became the process through which linguistic meaning was constructed (Besnier 1994).

One problem with education in Cameroon is that policies seem to be guided more by external factors than by local reality. The political desire to maintain a dwindling unification between the Anglophone and Francophone provinces was the basis for the decision to institute bilingualism at all levels of education. Yet very little was done to prepare the educational community for this. The pedagogic structure, set up in the ministry of education to monitor the implementation of this decision was, and always has been, under the pressure of a political and administrative structure that considers nothing but the political ambitions governing this policy. The result of such pressure is most often felt by the practitioner in the field, who is expected to produce the best possible results from students who are clearly not admitted on the basis of language proficiency, but on their overall performance in an examination that involves other subjects.

The frustration of not being able to shape decisions involving the teacher's practice and reality can be very demotivating. At the same time, being able to mediate between the dictates of administrative and political will on the one hand, and the constraints of local reality on the other, are very challenging. It takes a positive vision of one's profession and a hope that things will have to eventually change for a teacher working under such conditions to move on. For me, the challenge lay more in attaining examination objectives, that is, having my trainees succeed in a final examination that unfortunately tested language proficiency rather than methodological proficiency. There was a practical phase of the final certificate examination, whose evaluation depended not on the ELT expert, but on whoever was observing the trainee and this was most likely to be a non-ELT person.

Conscious of this shortcoming of the examination, and being unable to influence things, I became very frustrated, as I think most practitioners facing this situation would. Yet the power relationship within professions in Africa seems to persist in subjugating practitioners to the will of administrators, who in turn depend on their political overlords for the decisions they take. I do not intend to castigate politicians for meddling with educational matters; rather, I argue for an educational system guided by the political vision of a society but shaped by both the mediators (practitioners) and the beneficiaries (learners). This may seem too revolutionary, yet I believe that although it may be radical to get learners involved in making decisions on content and methodology, a sound understanding of their needs and expectations, of their abilities and difficulties, of the realities within which they operate is very important in taking decisions that will affect their education. Such understanding as I have described above has unfortunately not been considered by politicians so far and the almost complete absence of empirical research (apart from student's dissertations which serve shelves rather than minds) is a pitfall.

So far, I have concerned myself with explaining some of the difficulties that young professionals are likely to face without showing how these difficulties may impact not only on the pedagogy they adopt but also on their vision. I am happy to have started my career as a teacher trainer in a particularly challenging cultural and administrative context. Being compelled to develop a pedagogy that responded to the immediate context was an interesting challenge for me as it gave me an insight into the complexities of my profession. Many young teachers in Cameroon seek to work in urban areas where conditions are favourable for them. The result is that urban state schools become overstaffed with teachers working very few hours a week, while schools in rural areas suffer from a shortage of teachers. This means that the few teachers in rural schools have to work long hours, a situation which, although not financially rewarding, can be professionally helpful in that it gives young professionals greater opportunity to develop an affinity with classroom reality in a way their urban counterparts will not. Furthermore, the material limitations of rural communities necessitate the development of pedagogic creativity.

My students did not have textbooks, not just because they could not afford them, but also because the textbooks prescribed for their training were not available. My response was to develop community-based content that both reflected their realities and provided the language support they needed. This would not have been necessary if my learners had had all the prescribed material. What is more, in relying on resources that were available to them like folktales, local history and news, and making use of locally produced instructional materials, I was able to involve my trainees in developing as professionals in a way that was denied to me by my elitist and theoretical training. My immediate circumstances forced me, as it were, to reject the status quo and to look to local resources; drawing on these in a discourse of involvement with my trainees, a discourse that was symmetrical rather than asymmetrical and engaged rather than detached. In this way, I discovered a kind of therapy for the frustration of not being able to change decisions which I considered alien and inappropriate to my specific context, and at the same time I was able to bridge the gap created by cultural constraints and the power differential between me and my trainees.

I have explained the helplessness of practitioners when it comes to matters of educational policy and decision-making, but I think there is a nuance to this depending on the way a practitioner develops. As a novice trainer, I would have relied more on the advice of more experienced trainers had I not been working in a rural community where very few of my colleagues would want to work. However, I was alone and therefore had to grapple with my own reality. In Cameroon, the vast majority of teachers go through their professional life without being visited by a pedagogic advisor and this kind of isolation can be very detrimental to their professional development. Yet, its particular value lies in the fact that teachers with a positive vision of their profession are plunged into the realities and limitations of the educational system, stimulating creativity that can eventually build confidence in the teacher and positive attitudes in learners. In my case, the early experiences in my first school built my self-confidence and imbued me with a vision of teaching not as a challenge, but as achievement, and this vision of my profession came to govern my practice.

A shift in perspective

In 2004, after seven years of teacher training in the far north province and a further year in the south of Cameroon, I was appointed to the Ministry of National Education to work as National Pedagogic Inspector in Charge of Bilingualism. This marked the beginning of a rather more difficult challenge for a number of reasons. First, it is unusual in Cameroon – and I guess elsewhere – to move from the classroom to a position of policy-making without passing through the intermediary stages of divisional advisor and provincial inspector of pedagogy, which are necessary for preparing prospective policy-makers for their administrative and pedagogic roles. The second difficulty derives from a cultural issue I raised earlier about power relationships being governed by stereotypes about age. Being one of the youngest inspectors in the Ministry, I had to face the challenge of making my voice heard without overtly trying to overturn a traditional way of approaching the role which some of my colleagues had cherished for years.

A third challenge, much like the second, came from the new power relationship with colleagues, teachers of my generation and those with several years of teaching experience, some of whom had been my own teachers. Fortunately, my early experiences as a novice teacher trainer had, in a way, imbued me with the tools for dissipating any form of discourse collision that could emanate from the complex power differential created by age and authority. My wish to bring about change in the Ministry without undermining my older colleagues was fulfilled by engaging with them in terms of their discourse practices while trying to change these along the lines I have described in the last section, and this involved a profound awareness of, and sensitivity to, my own discourse acts that would appear as face threatening. (For an insightful discussion of how participants in multi-party interactions negotiate the need to present a positive self-image and the need to preserve the network to which they belong, see Diamond 1996, chs 3 and 4.)

In order to explain how the experiences I have described above helped me overcome these challenges, I shall describe my role in the area of English language teaching in French-medium schools. As National Pedagogic Inspector, my duty (with my other colleagues in the same inspectorate) was to define and monitor government policy in terms of the teaching of English in French-medium Nursery, Primary and Teacher Training colleges in Cameroon. I design syllabuses, monitor methodology and control end-of-course certificate examinations for English at the three levels. This entails a sound understanding of the socio-political and cultural vision of the country as well as the psychological and linguistic variables upon which language content and methodology are constructed. What is more, monitoring the implementation of policy involves developing and implementing effective parameters for pedagogic supervision of teachers and teacher trainers.

My first worry was caused by the title 'inspector', which gives the role a policing perspective and creates a gap between teachers and myself. To bridge this gap, therefore, I needed to create a working relationship that minimises any traces of administrative superiority and work with teachers on a horizontal basis. But in an educational culture where bureaucratic complexities tend to widen the gap between pedagogic authorities and the classroom practitioners, it becomes difficult to convince teachers that such an approach is not ensnaring for them. One challenge I have sought to overcome in the exercise of my duties, and which draws from my own early professional experiences, has been that of seeking to establish asymmetrical relationships that minimise the suspicion arising from encounters with unfamiliar discourse. In many situations, I have tried to negotiate my way into classrooms, but I understand that teachers admit me more out of a feeling of fear than from their own free will. The tendency for pedagogic supervisors to portray themselves, or be portrayed, as dictators of a pedagogy, both at odds with the cognitive needs of the teacher and alien to the immediate realities of teaching, has only helped to complicate the implementation of some education-related policies. Teachers are more likely to accept pedagogic innovation when it comes from amongst their own peer group rather than from hierarchy, an understandable response in a situation where such innovation is most often decided at a higher level and imposed with no regard to their realities.

In this respect, it is obvious that a change of attitude and approach to pedagogic inspection or supervision is necessary if we are to create a harmonious relationship between the 'governor' and the 'governed'. To obliterate the remnants of an educational culture built on strong and sometimes pervasive power relationships, positive change has to come from the top, but not as theory or policy, rather as practice that is accompanied by an enabling psychological and institutional environment that is capable of dissipating the likelihood of sanctions for pedagogic practices which, although inconsistent with bureaucratic prescriptions, are most often appropriate to the immediate context of the teacher.

Forming new bonds

One way in which I have been able to achieve a harmonious working relationship with teachers is through my involvement in the Cameroon English Language and Literature Teachers' Association (CAMELTA). Created in April 2001 as an umbrella association to the already existing provincial associations of English language teachers, CAMELTA aimed to:

- create a forum for the promotion of ever more teacher-friendly ELT
- create a forum for scholarship or learning
- generate networking and sharing of skills in best practice and new approaches in ELT
- stay trade union free and focus on professionalisation of ELT in our country, through in-service training of members.

Joining the association in 2001 in the far north province where I started my profession as teacher-trainer, I discovered that it provided a forum for teachers to share good practice as well as console one another on the frustrations imposed upon them by the more powerful bureaucrats. In 2003 I became president of the far north provincial chapter where I had the opportunity of networking with other provincial chapters and leading delegates to the national congress. The national congress gave me my first opportunity to interact with other members, some of whom were national pedagogic inspectors on an informal basis, as it was important – if the association was to grow – to create such an environment that would eliminate power relationships that were detrimental to the initial motives behind the creation of the association. In 2005, I was voted into the office of Secretary General and Chief Convenor of CAMELTA, thus placing me in a position to influence the destiny of the association.

Amongst other things, my stewardship within the association has not only helped bridge the gap between teachers and myself, but has helped me understand the wide range of difficulties teachers encounter in their day-to-day practice and such understanding is useful in shaping my vision of pedagogic supervision and inspection. As a community of practice (Wenger 1998), CAMELTA focuses on what members do: the practices or activities that indicate that they belong to the group. These practices or activities typically involve many aspects of shared behaviour, including discourse features and interaction patterns (Holmes and Meyerhoff 1999). Because CAMELTA provides a forum where pedagogic inspectors and teachers interact on an informal basis, because they all share the same goal of promoting quality ELT nationwide, and because teachers have the opportunity to take control of their organisation and share best practices, we have been able to establish the kind of relationship I have suggested above.

One way in which I have observed the importance of my role in the association vis-à-vis teachers is that while non-members of the association still tend to approach me from a power-inferior standpoint, members of CAMELTA do not feel inhibited working with me. Having encouraged teachers to invite me into their classrooms as a way of dissolving the face threatening effects of the intruding presence of a pedagogic inspector in their classes, I have observed that I am more likely to be invited by CAMELTA members than by non-CAMELTA members. By working with teachers through the shared aims of the association in a discourse community (see Swales 1990, notably ch. 2, for an illuminating discussion on the conceptualisation and operationalisation of speech and discourse communities), the barriers described above are overcome and the teachers themselves invite engagement. It is clear to me now that teachers' associations and professional networking can play an important role in building a professional community where practitioners and decision-makers can interact on the same grounds for the same purpose and that this, in turn, can bridge the gap that hierarchical stereotypes have placed between them.

Exploring new discourse

Late in 2005, I was invited to co-author a course book for English in French-medium nursery schools, an invitation which was followed a few months later in the next year by another one to write a course book for French-medium primary schools. The first, now in use in French-medium nursery schools was a kind of initiation into a new area of professional development which I came to see as exciting because it offered me another opportunity to put my experience and understanding of classroom reality into practice through the material I was to produce. Course-book writing, for me, involved a thorough analysis and understanding of the demands of the syllabus as well as a probing into classroom reality. While I can claim that I had control over the former, having participated in developing the syllabus for English at this level, it is hard to say if my proficiency in the latter was anything to rely on. Having had the chance to interact with and observe nursery school classes in different parts of the country and having had the opportunity of 'fiddling' with the nursery classroom as part of my training strategy in my early days as a trainer, I had something to draw on, but I had no access to empirical research in order to inform my decisions. The course book for primary schools has taken a longer time because of the many levels involved and this has been the most exciting writing experience for me because it has been undertaken through a different period of theoretical awareness in my life.

In 2006–7 I won a Hornby Scholarship to study in the United Kingdom for an MA in language teaching for young learners. This stage of my professional development is certainly the most rewarding, not because it offers me any practical solutions to the challenges I have described all through this chapter, but because it has armed me with a wide variety of theoretical and philosophical perspectives underpinning pedagogic practice. Blending my studies with my ongoing exploration of a new discourse through course-book writing, I have been able to involve myself in new discourse encounters which will inevitably be introduced into my work as pedagogic inspector and member of a community of practice to expand its horizons. I shall not go into the importance to my experience as a teacher of each course I have taken, but will broadly show how my vision of my roles as teacher trainer, pedagogic inspector and course-book writer have been greatly influenced by a new vision of myself as researcher engaged in a discourse of discovery.

Exposure to a vast amount of literature in ELT practice and theory has helped me revisit my practice in the light of research and this has lent credibility to what I would hitherto have considered as simply intuitive. An awareness that one's practice can be supported by empirical evidence provides a kind of confidence and assertiveness that helps not only to influence practitioners' discourse, but also impacts positively on their professional development. From a methodological stance, my exposure to sociocultural psychology and collaborative learning, as well as my reflections on learner autonomy and motivation (also see Kuchah and Smith, forthcoming) have enriched my understanding of the theoretical constructs behind most of my practice. My initial training built on constructivist psychology drawing on a Piagetian perspective and Bloom's taxonomy but, as I mentioned earlier, whatever psychological concepts I learnt at that time were relevant only for my examinations.

With a professional experience of over ten years, spent through different stages of professional development, it is easier for me to revisit my practice in the light of theory and ascertain the relevance or irrelevance of certain aspects of the theory to my particular context. This would not have been possible in my initial training because there was no practical experience to act as counterbalance to the theoretical input I got and what is more, such input, acquired in an exam-oriented context, could hardly have been relevant to me as it is today. I strongly believe in a theory–practice relationship of checks and balances, with theory directing practice, and practice in turn correcting theory and providing a framework for further research.

This belief in the relationship between theory and practice, emanating from my in-depth encounter with theory at this stage of my professional development, has also helped to redirect my vision of my role as materials designer and pedagogic inspector not just from a psychological and methodological perspective, but also from a linguistic perspective. My approach to materials design has been influenced by an understanding of the strengths and limitations of CLT, which in my working context has been the subject of multiple interpretations, some of which have led to a *laisser-faire* approach to language teaching by some teachers. Course-book writing, for me, now involves not only a mastery of curriculum demands, but a firm theoretical understanding of both psychological and linguistic considerations governing language teaching and learning. Here, my courses in Applied Linguistics and Professional Practice have been particularly helpful in that they have provided me with the different research perspectives that govern language teaching pedagogy. My approach to materials design is thus not just directed by my understanding of the syllabus, but most especially by an awareness of interactive activities necessary for sustaining both the attention and interest of young learners and this awareness is a result of my exposure to empirical research. The debate about authentic versus non authentic texts, although not decisively useful to me, has helped me revisit the kind of material I design in the light of their relevance and familiarity to my target learners' day-to-day realities.

Conclusion

Professional development in sub-Saharan Africa is as challenging as elsewhere, but it has the added burden of a demanding socio-political and economic structure which imposes patterns of discourse on teachers that may not genuinely reflect their immediate realities. The asymmetrical discourse of power, of the imposition of the status quo I have described above, reflects not only a postcolonial legacy but also a local culture of hierarchies which is difficult to challenge. The problem with such a top-down professional system is that apart from being insensitive to the cultures and traditions of teachers and learners, it severs the connection between decision-makers and practitioners, plunging the latter into an isolation that can be counter-productive to the attainment of pedagogic goals. Yet, as I have shown in this chapter, it can be challenged by connecting with local culture and reality (also see Kuchah and Smith, forthcoming), and by forming new discourse communities and communities of practice where shared goals can be negotiated in a symmetrical relationship.

The role of hierarchical structures should be to help teachers grow professionally, by creating an enabling psychological environment which takes into consideration teachers' responses to their changing environment and constantly provides them with a theoretical basis for reflecting on their practice. Africa needs to focus attention more on teacher education than on teacher training because to date, the latter, whether pre- or in-service, has been flawed by stereotypes of power relations and the inappropriate transplanting of a theoretical stance not fully relevant to the immediate realities and needs of the teacher.

I do not, by this, mean to undermine the important place of pre-service and in-service training in effecting the transmission of innovative practices; what I argue is that we need a change of attitude to pedagogic innovation and that such change does not come through mere training, but through education which involves developing a spirit not only of reflective practice but also of a harmonious relationship between, and respect for, the potentials and opinions of all stakeholders. The practitioner is at the very centre of the educational structure, mediating between the demands of a curriculum that does not always consider his or her opinion, and learners whose needs and interests are as varied as are their numbers. Pedagogic supervisors can play a vital role by avoiding practices that tend to victimise teachers rather than exhort them to develop their practice. A possible route to this is for decision-makers to engage with practitioners in a discourse of involvement, through communities of practice where, with exposure to new discourse, new bonds upon which new discourses will be explored can be formed.

Note

1. Because Africa is presently under the pressure of sociocultural and political change and each country faces realities that are both similar and distinct from other countries, what may be seen as general practice in one country is very likely to be inconsistent with the realities of another. This caution, I think, is worth observing when attempting to make generalisations about professional development and discourse patterns in a continent that is caught up between so many indigenous cultures, languages and beliefs, including Western cultures imposed through colonisation, on the one hand, and in the world vision of a people struggling for self identification on the other.

References

Besnier, N. 1994. 'Involvement in linguistic practice: an ethnographic appraisal'. *Journal of Pragmatics*, 22: 279–99.
Constitution of the Republic of Cameroon (1996).
Diamond, J. 1996. *Status and Power in Verbal Interaction*. Amsterdam: John Benjamins.
Fairclough, N. 1989. *Language and Power*. London: Longman.
Holmes, J. and Meyerhoff, M. 1999. 'The community of practice: Theories and methodologies in language and gender research'. *Language and Society*, 28: 173–83.
Kuchah, K. and Smith, R. C. (forthcoming) *Learner Autonomy: Practice in Difficult Circumstances*. Dublin: Authentik.
Law N°98/004 of 14 April 1998 to Lay Down Guidelines for Education in Cameroon.
Swales, J. M. 1990. *Genre Analysis*. Cambridge: Cambridge University Press.
Wenger, E. 1998. *Communities of Practice: Learning, Meaning and Identity*. Cambridge: Cambridge University Press.

11
Becoming a Writer: Community Membership and Discursive Literacy
Sue Wharton

Introduction

What does it mean for English Language teachers to write for publication? What difference does it make to how they see themselves and are seen by others, to the contribution they make to the profession? This chapter is in the section of the book entitled *Passing on the Knowledge*; this, then, is part of what language teachers do when they write. In this chapter I aim to explore in more detail both what is involved and what it feels like to people beginning to take on this role.

I need to be clear straight away about the kind of writing I will discuss. I am referring specifically to writing published in journals or magazines whose primary readership is other TESOL professionals. As I will discuss below, the range of such publications is wide and there is continuing debate about how to label and/or categorise the work reported therein – whether as research, as descriptions of practice, as practitioner research and so on. The choice of terms is important, connoting ongoing professional debates and battles about the status of different contributions and individuals (for a recent example, see Bartlett and Burton 2006). But in this chapter, with its focus on becoming a writer and the process of writing for publication, I want to remain as separate as possible from those arguments about how to label the product. Rather, I will look at the experiences of some teachers as they begin to write for publication, and consider the implications for other teachers wishing to do the same.

What do EL teachers write about?

Widdowson (2003) begins his edited collection on defining issues in ELT with the assertion: 'There is a great deal of distrust of theory among

218

English language teachers. They tend to see it as remote from their actual experience, an attempt to mystify common-sense practices by unnecessary abstraction' (2003: 1). On reading this I had several reactions. The first was simply to wonder whether Widdowson is right. His assertion suggests that teachers see theory and practice as dichotomous, a view that Clarke (1994) and others since (e.g., Delandshere 2004; Pring 2000; Roulston, Legette et al. 2005) have claimed is unhelpful for the development of the profession. EL teachers form a heterogeneous group, and many are more open-minded than this.

The second was to wonder what such an assertion implied about published writing by EL teachers. Presumably, teachers do not themselves write the 'theory' which they 'distrust'; Clarke (1994) specifically argues that overly-general theories, those which risk dysfunctionality because they do not take sufficient account of specific instances of implementation, are not written by teachers themselves. Could it be that teachers, connected by daily practice to the importance of specific instances of implementation, are in a particularly strong position to write professional accounts which do not fall into this trap?

Foster (1999), commenting on accounts generated by teacher research in the United Kingdom, states that the writing is predominantly 'personal descriptions of, or justifications for, their own practice; or accounts of their efforts to improve pupil achievement, or of their involvement in staff development activities' (1999: 383). For Foster, this is a criticism – he argues that such accounts fall short as research. For me, his comment is potentially valuable as revealing precisely those issues that may be of interest to practising teachers, both to investigate and to write about.

Previous research on articles published in the mid-1990s in *ELT Journal* (Edge and Wharton 2002; Wharton 1999) suggests that four major types of article can be found there. First, 'action' articles which detail procedures undertaken towards some goal. Their purpose seems to be to invite the reader to consider similar courses of action in appropriate circumstances. Second, 'rationale' articles which explain and argue the validity of a particular course of action or approach. Again, the purpose seems to be to suggest that the reader consider a similar approach. Third, 'analysis' articles which explore and interpret a situation. Their aim seems to be to increase understanding, without necessarily advocating specific action. And fourth, 'evaluation' articles, which evaluate work other than that of the authors, or approaches which are well known within the profession. Here, the purpose seems to be to convince readers of the validity of the perspective.

ELT Journal is only one publication of course. Many articles in it are written by teachers, and the journal specifically requests articles with a direct connection to practice. However, the article 'types' discussed above are also to be found in other publications. It is interesting to note some similarities between the topics of accounts which Foster criticises, and the topics which this well-known journal seems to value. And there is a link, of course, between preferred topics of investigation, preferred methods of investigation, and preferred styles of writing up. Like any profession TESOL has its own value criteria, and the types of articles written by teachers are one indication of the theory–practice connections which are held to be useful.

Where do EL teachers publish?

Commentators continue to lament the relative absence of work by practising teachers in refereed journals (e.g., Burns 2005; Hayes 1996; Reis-Jorge 2007). Egbert (2007) provides a review of refereed journals in TESOL. She discusses the perceived hierarchy of journals, the domination of some of them by writers in university positions, and the notion that a high rejection rate is a perceived indicator of quality. Such a situation can hardly encourage more teachers to write for publication and thereby redress the imbalance lamented.

Fortunately for those of us who wish to read accounts by practising teachers, there are a variety of outlets which do encourage such work. Certain refereed journals, such as *ELT Journal*, discussed above, publish work from people in various positions within the profession, including a high proportion of EL teachers. *Language Teaching Research* is another example, in this case with a specific section dedicated to practitioner research. EL teachers also publish in professional magazines such as *English Teaching Professional*; in conference proceedings, such as *Teachers Develop Teachers Research*; and, arguably increasingly, in edited collections. A recent, prize-winning example would be Edwards and Willis (2005). For a writer, then, there are important decisions to be made regarding the placing of a paper or the drafting of a paper with a particular outlet in mind. EL teachers, of course, are in a good position to analyse the texts which appear in different journals, and so make informed decisions about their own writing.

Who is the audience?

Our professional community is very diverse. The audience for EL teachers' writing includes other teachers, teacher educators, researchers, student teachers, experienced teachers undertaking higher studies, in rich and in poor countries and in state and private systems all around the world. Some TESOL publications target only a small section of the community, but others aim to have a broader appeal.

Whatever the particular remit of the journal, magazine or book, contributions within it have a pragmatic purpose in common: the writer has something to suggest to the readers (see Edge, this volume). Readers, in turn, are most likely to turn to literature of this type because they want to get something from it: ideas to implement in a language class, ideas to discuss among colleagues or to put forward in a teacher education setting, references for an academic assignment, or for their own intended writing for publication.

To write for a wide readership, with a wide variety of instrumental purposes for reading, is of course a challenging task – a writer needs to address both those readers who are familiar with the particular topic at hand, and those for whom it may be unknown territory. TESOL writers have developed some interesting discursive strategies for negotiating these demands, some creative uses of genre conventions in order to highlight the value of their contribution (Wharton 2006). Later in this chapter I will discuss some of these. But first, in the following section, I will look at some aspects of the challenge in more detail.

Some challenges in becoming a writer

Writing for professional publication means the creation of a text that stands alone as an expression of ideas; unlike when teaching, one cannot monitor the audience and adjust one's explanations accordingly. And yet, this text is simultaneously a discourse act through which writers situate themselves in their professional community. As Hyland points out, a writer must 'demonstrate that they not only have something new and worthwhile to say, but that they also have the professional credibility to address the topic as an insider' (2000: 63).

Harre's (1983) model of social and psychological development offers a useful resource for understanding of the mechanisms by which this position may be achieved. This model can be understood as throwing light on some of the intellectual and social processes which underly the act of writing in a professional community. Harre's cycle has four phases: appropriation, transformation, publication and conventionalisation. As they relate to professional writing, *appropriation* refers to the writer's familiarity with community knowledge in the area where they want to write. *Transformation* refers to their personal perspective on these issues, the interaction of knowledge, ideas and experience which allows the writer to develop a new contribution. *Publication* happens when a stand alone product is created, a text through which the contribution of the writer is brought into the public arena. *Conventionalisation* refers to the possible uptake of those ideas within the professional community; the text becomes a resource on which others can draw in their own processes of appropriation.

Published professional writing, at whatever length and in any publication recognised in the community, can be seen as an outcome of the cycle discussed above. For this reason, published writing tends to construct its author as an expert. By its existence it claims that the author has a contribution to offer.

Becoming a published writer, then, involves a teacher in an extra dimension of professional identity, an additional layer of expertise. The teacher must have confidence in their ideas and their authority to articulate them. Equally, they need sensitivity towards potential readers' own ideas, authority and contextual knowledge. In order to successfully inhabit and project this complex identity, a teacher can benefit from developing their own discursive literacy: conscious, critical awareness of the forms and functions of genres through which the TESOL community debates its ideas. It is to this dimension that I now turn.

Developing discursive literacy

All of my own work in this area, developing my own discursive literacy and seeking to facilitate such development with others, has been based on Hoey's (1983, 2001) work on semantic relations. It has involved reference to that textual pattern which is globally termed 'Problem–Solution' to understand and explore the semantics, pragmatics and purposeful variation of professional writing in TESOL. Much of the work was in collaboration with Julian Edge, and he describes the ideas behind it in his chapter in this volume – I will not discuss them in detail here. Rather, I want to focus on the perspectives of some EL teachers as they worked with a semantic patterning approach to the development of their own TESOL writing intended for publication. I will examine the challenges they saw themselves as facing, and at some possible strategies for overcoming these.

The writers quoted in this section were, at the time of our work together, practising EL teachers and also participants on, or graduates of, a distance learning Masters in TESOL run from Aston University, United Kingdom. None were previously published writers, and all had material which they wished to develop as writing to submit for publication. I offered them a set of materials designed to facilitate conscious awareness of some of the goals of TESOL writing and, via a semantic patterning approach, to consider how well their own writing was working. The comments below are taken from email, telephone or face-to-face discussions between us. I use them here to illustrate some of the major topics which emerged.

Authority and insecurity

A first challenge is the sense of insecurity which I think many of us would identify with as we embark on a new project. When contemplating an addition to our professional role, most of us would be worried about how we were going to fit it in alongside all the other activities. We may also feel concerned that we do not know quite how to do the activity itself, and are likely to waste time trying to find out. A teacher in Mexico commented:

> I think most people would appreciate more authoritative guidance as this often signifies for most a kind of security...

There is of course no way of avoiding this sense of insecurity, and so the feeling itself cannot be 'overcome'. However, there may be ways of continuing to write despite the anxieties, and gain a sense of authority. For example, a teacher in Japan commented:

> My initial impression, personally, is a drastic change in my own outlook to publication. When I [began this work] I felt...unable to speak 'expertly'. But, after getting a few more assignments under my belt, working with the professionals at Aston, meeting editors, etc, I feel my research now contains enough 'new' that I can approach it as somewhat of an expert.

This teacher felt insecure about his authority; however, he continued to write and to network with more experienced writers. In this way, he developed a sense of how to construct his own authority.

A teacher working in the United Kingdom talked about the value of studying the kinds of texts he wanted to write. Specifically, he found that the semantic patterning approach we were using helped him to understand that the idea of a new contribution was a textual concept:

> While I suppose I knew that newness and interestingness were in part artefacts of discourse, [this work] gives me new confidence to approach writing as a 'member of a discourse community'.

A teacher working in Italy talked about her increasing awareness of different publications identifying themselves as addressing different audiences within the TESOL community. Having seen her writing as identifying, in Hoey's terms, a Problem and a Solution, she considered whether these were likely to be already familiar to an intended audience, or not. This assessment allowed he to make strategic decisions about where to place her work. She chose to position it in a context where she felt that neither Problem nor Solution would be entirely familiar, and so where the work would function most effectively as a contribution.

Getting the reader–writer relationship right

A further set of challenges concerns the delicate discourse task of setting up a Problem to which a particular piece of writing can respond. Problems must belong to somebody, and sensitivity is needed when deciding where to attribute them within the community. A teacher working in Turkey commented on the following section of his draft text:

> This [difficult state of affairs] is generally because language teachers feel tightly bound by their traditional classroom role as pedagogues – 'providers of information on formal rules'. They do not generally see themselves as 'reflective practitioners' who might benefit from observing other teachers and, in turn, being observed.

He decided to revise the second sentence: 'They do not generally see themselves as reflective practitioners…', as he decided that in its present form it could alienate those teachers who *do* see themselves as reflective practitioners.

A similar point was made by a teacher working in Italy, already mentioned above. She spoke of her wish to indicate that she was working in an under-researched area without disparaging existing research. She specifically substituted 'no attempt' with 'little attempt' in the following paragraph:

> In spite of growing recognition of the importance of leaner initiative, there has been little attempt either to define what it means or to analyse the ways in which this initiative is expressed and the effects it may have on classroom interaction.

Many TESOL articles make use of a Problem–Solution pattern but, as we have seen, employment of this pattern obliges the writer to perform a delicate balancing act in their role as a member of a professional community. They need to argue that their chosen focus is an issue worthy of attention, without disparaging their colleagues who may not have demonstrated awareness of this.

How much to tell

This topic relates specifically to action texts, those which discuss a course of action or a set of procedures. A teacher working in Spain raised questions regarding the extent to which a published text should include only positive evaluation of such procedures:

> [In the examples we have looked at] All the evaluations mentioned are positive, no negative aspects are mentioned → does this mean that no negative aspects (i.e. objections) should be included?

She herself preferred to write an evaluation which did include some more negative aspects, as follows:

> Although this two part system, together with the more quantitative aspects mentioned in passing, allowed me to reach a satisfactory assessment of the trainees' work in the course as reflected in their diaries, there are at least two points where this method caused problems, and which I would like to point out briefly.
>
> (Elaborates the two points in one paragraph each)

She argued that the negative evaluation was important in moderating the claims made in the text, in order to avoid presenting herself as having found a cure-all solution in a particular context. On the other hand, she wondered whether the claim for expertise, an important part of writing for publication, could still be discoursally managed with negative evaluation included: 'perhaps the form you give this evaluation has to be different?' These topics bring to mind those of Foster (1999) regarding teachers' writing as 'accounts of their efforts to improve pupil achievement' (1999: 383). The publication of accounts which include negative as well as positive outcomes provides a valuable learning opportunity for readers; a vicarious understanding of obstacles as well as successes. Writers of action-focused texts can consider the extent to which such information and reflection will be of value to their readers, and include it accordingly.

The topics discussed in this section are some of the most frequent ones that emerged in the work undertaken at Aston. There is of course no sense in which they are offered as an exhaustive discussion of issues facing new writers. However, they do provide some specific insights into the thought processes of teachers beginning to write for publication, some challenges which they faced and how they worked to solve them.

Teaching and writing as two aspects of a professional identity

So far in this chapter I have discussed some of the opportunities for writing for publication in TESOL, I have brought in the thoughts of some EL teachers as they made this move in their own professional lives, thereby discussing some of the difficulties involved, and some of the strategies particular writers have used. However, the title of the chapter is 'becoming a writer' and that, of course, goes beyond the production and publication of an individual text. The shift in identity required to find a published voice may also resonate with, and contribute to, a more profound professional development, if one becomes the kind of teacher who contributes regularly with one's writing.

The dual identity of teacher and writer is complex for a variety of reasons. Some of these are practical. Reis Jorge (2007) quotes the views of teachers who are interested in research and investigation but tend to feel that the demands of writing this up would be unmanageable in terms of time. Similar comments were made to me during the project at Aston. For example, a teacher in Mexico mentioned:

> I think people in general, not just us, always look for a quick recipe of how to do things as apart from the security aspect it gives the impression of being a time saver and here we are always fighting against the clock.

Such time constraints are obviously very real and lead to a question: What situations might function as opportunities in which these practical constraints could be lessened or overcome?

One possibility is the undertaking of further study. As I mentioned above, the work done at Aston was with teachers studying at Masters level and for these people, the course itself seemed to have opened some opportunities to start writing for publication (Wharton 2002). However, there is also a downside to an academic study situation. Winter (1998) points out that such a situation adds yet a further complication to the identity of the teacher. Winter suggests that professional practitioners taking university courses see themselves in that context as students, depending on the course or the tutors to validate their ideas. This, of course, may not be conducive to developing a position of expertise from which to write in one's own voice.

In Reis Jorge's (2007) study referred to above, some informants specifically link the issue of whether they are likely to write with the question of the job they might have in future. They seem to see a school-based job as making writing impossible due to lack of time, and a university-based job as obliging them to write. Reis Jorge quotes one informant as follows:

> If we go back to schools the possibility ... of carrying out any research I think will be very slim. However, if you are posted to teacher training then I think it's one of the criteria, while being a lecturer where you need to carry out research, do publications.
>
> (2007: 411)

This teacher raises a key issue: the difference between writing for publication as part of one's job description, and writing for publication when this is not an obligatory activity. I personally can identify very easily with the position described. I became interested in the idea of writing for publication after a few years of teaching, I would even go so far as to say that I formed an aspiration to do so. But I did not take any concrete steps towards achieving my ambition until I moved to a professional position in which I saw it as part of the job description.

I must say though that I now find my reticence regrettable – or rather, I feel it would be regrettable for the profession if all teachers felt the same way. We would lose a lot of resources. In the final sections of this chapter, I want to put forward an argument about why I think it is particularly valuable for our profession as a whole to have access to the writing of teachers for whom publication is not an obligatory activity.

The importance of teachers becoming writers

Chouliaraki and Fairclough (1999) argue that any practice, such as teaching, includes as an inherent part of itself representations of what those involved in the practice think they are doing. Published professional writing is, of course, one place in which such representations can be found.

Such reflexivity would seem a self-evidently good thing, but Chouliaraki and Fairclough point out that it can lead to 'theoretical practices', that is, the specialised production of knowledge about practices. Professional writing is an obvious example of a theoretical practice; as such it has its own coherence, internal logic and skills to be learned, which are different from skills in the practice itself. This idea perhaps links with the comment of Widdowson at the beginning of the chapter – perhaps 'teachers' distrust of theory' is a matter of practitioners distrusting those they see as having expertise in a theoretical practice rather than in the practice itself. Chouliaraki and Fairclough also point out that theoretical practices easily become tools of power and domination, so that those who possess expertise are in a better position than those who 'just' practice. This is another aspect of the dichotomy referred to by Clarke.

It seems to me that such points are the strongest possible argument in favour of the dual identity of teacher and published writer. Individuals who are willing to take on this dual identity provide the profession with an invaluable resource, that is, an accumulation of accounts written by people with expertise *both* in the practice of EL teaching, *and* in the theoretical practice of writing about it.

In my current role, teaching on a Masters course in TESOL, I find it essential to be able to draw on such accounts and to encourage those taking the course to read them. For example, if teaching syllabus design, I am just as likely to recommend a book like Graves (1996), which is a variety of accounts from teachers of their course design projects, as I am to recommend any writing on principles of course design. Were it not for the existence of these accounts, the learning experience of teachers-in-education would be considerably diminished.

Some might argue that the voices of practising teachers can be brought into the published discourse of the profession even when they do not author their own accounts. For example Hayes's (1996) article has the powerful title 'Prioritizing voice over vision: Reaffirming the centrality of the teacher in ESOL research'. He states that one of the purposes of his article is 'the sponsoring of teachers' voices from a non-western context' (ibid.: 174). The voices are indubitably present, via detailed quotation from interviews – but the author of the paper is not a teacher himself; he is a university-based teacher educator. Reis-Jorge (2007) comments with regret that 'much of what has been written about teacher-research has been by academic researchers and educational theorists' (2007: 403). His own paper certainly includes the voices of teachers, again via detailed extracts from interviews; but again, these teachers are data sources, not authors.

I do not wish to criticise the work of Hayes or Reis Jorge for including the voices of teachers as data – this is in itself a perfectly valid procedure (and one which I have used in this chapter). But it is not, in my view, an alternative way of giving those voices weight in the debates and developments of the profession: appearing as data in someone else's publication is hardly the same as appearing as the author in one's own.

As Frow (1998) points out, speaking for oneself is an indicator of power, whereas being spoken for by others suggests relative powerlessness. He quotes Alcoff: 'the practice of privileged persons speaking for or on behalf of less privileged persons has actually resulted (in many cases) in increasing or reinforcing the oppression of the group spoken for' (1991: 6–7, cited in Frow 1998). When teachers become writers, they not only pass on their knowledge and perspectives, they also claim the activity of writing for professional publication as part of their identity – they do not rely on others to speak for them, but speak for themselves.

Managing the dual identity

I have described above how teaching and writing can be seen as dual aspects of a professional identity. In this penultimate section, I will recap on some of the comments of teachers discussed earlier, and consider the implications for a teacher becoming a writer. The comments relate both to the management of one's professional life, and to the management of particular texts.

From the first perspective, writers discussed the importance of networking with experienced colleagues and of becoming familiar with potential outlets for writing. These strategies can help new writers make good decisions about where to place their work. From the second perspective, it seems that awareness of the Problem–Solution text pattern and its function in this genre can help a writer to shape an emerging article. Such awareness can help a writer consider appropriate language choices when positioning themselves vis-à-vis their audience, and when making claims for the significance of the work.

Some final thoughts

In writing this chapter, I have had a problem with pronouns. I have referred to EL teachers, and EL teachers engaging in writing for publication, as 'they'. For those readers who identify themselves as EL teachers, the implied pronoun will be 'you'. And in many ways, I would have preferred to use 'we'. My strongest sense of professional identity is as a teacher, and my published writing is about teaching, about language, or about both. So why did I not feel that I could claim the pronoun 'we'? I think there are three reasons.

Firstly, because this is a book about EL teaching and I myself am not an EL teacher. Secondly, because this chapter focuses on becoming a writer, once one has made that step, one cannot pretend to be still thinking about it. Thirdly, perhaps most importantly, because of the distinction made above between teachers – usually in Higher Education – for whom writing for publication is a job requirement, and teachers in a far wider variety of institutions who undertake the task and the role without being given official time or space for it. It is primarily this latter group of teachers that this chapter is about, and to whom it is addressed.

References

Alcoff, L. 1991. 'The problem of speaking for others'. *Cultural Critique*, 20: 5–32.

Bartlett, S. and Burton, S. 2006. 'Practitioner research or descriptions of classroom practice? A discussion of teachers investigating their classrooms'. *Educational Action Research*, 14(3): 395–405.

Burns, A. 2005. 'Action research: An evolving paradigm?' *Language Teaching*, 38(2): 57–74.

Chouliaraki, L. and Fairclough, N. 1999. *Discourse in Late Modernity*. Edinburgh: Edinburgh University Press.

Clarke, M. 1994. 'The dysfunctions of the theory/practice discourse'. *TESOL Quarterly*, 29(1): 9–26.

Delandshere, G. 2004. 'The moral, social and political responsibility of educational researchers: Resisting the current quest for certainty'. *International Journal of Educational Research*, 41(3): 237–56.

Edge, J. and Wharton, S. 2002. 'Genre teaching: The struggle for diversity in unity. In K. Miller and P. Thompson (eds), *Unity and Diversity in Language Use*. London: BAAL/Continuum, pp. 22–38.

Edwards, C. and Willis, J. 2005. *Teachers Exploring Tasks in English Language Teaching*. Basingstoke: Palgrave Macmillan.

Egbert, J. 2007. 'Quality analysis of journals in TESOL and Applied Linguistics'. *TESOL Quarterly*, 41(1): 157–71.

Foster, P. 1999. 'Never mind the quality, feel the impact: A methodological assessment of teacher research sponsored by the teacher training agency'. *British Journal of Educational Studies*, 41(4): 380–98.

Frow, J. 1998. 'Economies of value'. In D. Bennett (ed.), *Multicultural States: Rethinking Difference and Identity*. London: Routledge, pp. 53–68.

Graves, K. (ed.) 1996. *Teachers as Course Developers*. Cambridge: Cambridge University Press.

Harre, R. 1983. *Personal Being*. Oxford: Blackwell.

Hayes, D. 1996. 'Prioritising "voice" over "vision": Reaffirming the centrality of the teacher in ESOL research'. *System*, 24(2): 173–86.

Hoey, M. 1983. *On the Surface of Discourse*. London: George Allen Unwin.

Hoey, M. 2001. *Textual Interaction: An Introduction to Written Discourse Analysis*. London: Routledge.

Hyland, K. 2000. *Disciplinary Discourses: Social Interactions in Academic Writing*. London: Longman.

Pring, R. 2000. *Philosophy of Educational Research*. London: Continuum.

Reis Jorge, J. 2007. 'Teachers' conceptions of teacher research and self-perceptions as enquiring practitioners: A longitudinal case study. *Teaching and Teacher Education*, 23(2): 402–17.

Roulston, K., Legette, R., DeLoach, M., and Pittman, C. B. 2005. 'What is research for teacher researchers?' *Educational Action Research*, 13(2): 169–90.

Wharton, S. 1999. 'From postgraduate student to published writer: Discourse variation and development in TESOL'. Unpublished PhD thesis. Aston University, UK.

Wharton, S. 2002. 'Writing from a context: Course assignments and professional development'. In M. Graal (ed.), *Changing Contexts for Teaching and Learning: Selected Papers of the 8th Annual Writing Development in Higher Education Conference*. Leicester: University of Leicester, pp. 153–68.

Wharton, S. 2006. 'Given and new in TESOL texts: The management of community consensus and individual innovation'. *IRAL*, 44(1): 23–48.

Widdowson, H. (ed.). 2003. *Defining Issues in English Language Teaching*. Oxford: Oxford University Press.

Winter, R. 1998. 'Managers, spectators and citizens: Where does "theory" come from in action research?' *Educational Action Research*, 6(3): 361–76.

12
Discourses in Search of Coherence: An Autobiographical Perspective

Julian Edge

Introduction

The invitation to contribute to this collection offers me an opportunity to bring together what I regard as significant discourse parameters of a working life in order to see if, from the perspective of now, any useful constructs emerge. In line with the design of the collection, I also reflect on the process of passing knowledge on – an experience that involves as much getting as giving.

The three major discourse-oriented strands of my work in TESOL arise from the study of semantic patterning in text (Edge 1989; Edge and Wharton 2002), the praxis-related discourse of action research (Edge 1994, 2001; Edge and Richards 1993) and the non-judgemental discourse of Cooperative Development (Edge 1992, 2002, 2006). I review them in that sequence.

Semantic patterning

It is easy to identify Michael Hoey's (1983) *On the Surface of Discourse*, as the single most important text in giving my work a research orientation. I had previously worked (for four years at the University of Alexandria and three years at the University of Istanbul) on the pre-service training and education of the next generation of English language teachers. One thing I knew was that, while I could set my undergraduate students useful articles to read from, for example, *Modern English Teacher, Forum* and *ELT Journal*, articles that I thought they would be able to understand, nothing very much ever came of it. From Hoey, I learned (among much else) two central points.

First, the ubiquity of the Problem/Solution pattern (Situation, Problem, Response, Evaluation) in organising texts in many genres. Second, the even more powerful status of this pattern as a culturally specified cognitive framework of interpretation:

> If the claim is correct that there are an infinite number of discourse pattern possibilities, then the prevalence of any recurring patterns must be explained. This can be done if they are seen as culturally approved patterns which reflect (and perhaps to some degree influence) the Western world's concern with problem-solving and classification.
>
> (Hoey 1983: 178)

Toward my professional ends, it became useful to point out to teacher trainees how frequently the pattern of an article about ELT methodology in the journals we were using was one of:

Situation:	This is where I work.
Problem:	This is a problem we have.
Response:	This is how I respond to that problem.
Evaluation:	Here is my evidence for why I say that this is a good idea.

The second key to comprehension for students such as mine was to recognise the pragmatic significance of what they were reading. That is, to realise that these articles were not merely comprehension passages on which one had to answer questions, although their previous experience might well have led them to believe that this was the goal of reading in English, just as their previous educational experience of interaction with texts in their own language might well have led them to believe that the memorisation of the content of those texts was the fundamental educational purpose of this activity. They had to realise, centrally, that the Response section of the articles had the pragmatic force of Suggest. Whether the Response was expressed as a past-tense narrative, or a simple-present description of procedures, or a string of imperatives, the function was the same: *the writer was making suggestions for what the reader should do.* Everything else in the article served the writer's purpose of getting that suggestion across as persuasively as possible; that was what the article was *for.* The reader's responsibility was to evaluate the worth of that suggestion from the perspective of the reader's own contextualised, professional action. The reader's responsibility to evaluate and act, rather than only understand and remember, was crucial (Edge 1989).

The usefulness of this semantic/pragmatic patterning in my teaching has not diminished, although the students that I now mostly work with have changed in character from undergraduate to postgraduate.

What I have learned since shifting, in 1986, into the teaching of master's students in British universities, is how powerful the making explicit of these basic elements is in supporting the professional and academic development of course participants whose level of English language ability is very highly developed, including native speakers of dominant dialects such as British and US English. This power has been, and continues to be, used to support not only course participants' professional reading, but also their writing (Wharton 2001). Master's students have been encouraged to use the pattern to shape the assignments that they write and further research (Edge and Wharton 2003; Wharton 1999; Wharton, this volume) has used the same approach to explore differences between assignments and published articles, at the same time developing materials to help novice researchers make this transition from one discourse community into another.

The pedagogic adaptation of Hoey's analytical model has led to one change that I continue to find stimulating, from an SPRE to an SFRE pattern, replacing Problem with Focus (cf. Hoey 2001). This idea of Focus captures the reading or writing move that one is looking to encourage, as well as opening up the space between the initial Situation and the final Evaluation to a broader, and I believe more powerful, if abstract, conceptualisation.

With Situation and Evaluation the basic building blocks of text (Winter 1986), I have come to think of what occurs between them as the *central dynamic* (see also Wharton, in press) of the text. This central dynamic might take many forms, the range of which can only actually be limited by the nature of the initial Evaluation of Situation.

To put that in more concrete terms, let us imagine a situation in which a textbook prescribed for a course cannot be delivered in time. Where I see a *problem* that requires the selection of a different title, you may see an *opportunity* to take teaching online, while Tom sees a *challenge* to rise to in terms of materials production, Dick sees a *resource* to exploit in the shape of an unpublished manuscript, and Harriett identifies the *goal* of improving ordering systems with the campus bookstore. While the nature of this Focus is determined by the initial Evaluation of the Situation, it is the Focus itself that defines the nature of the appropriate Response, and the Response, in turn, that defines what will count as appropriate evidence, or criteria, for subsequent Evaluation. We can expect that the nature of the affordance (van Lier 1996) identified as Focus will have an effect on the readiness, and effectiveness, of the Response that it evokes. Furthermore, if we acknowledge the importance (albeit under-investigated) of the emotional, as well as the rational, elements of teacher cognition (Borg 2006: 272; Zembylas 2005), we might also expect that a preference for a certain kind of action (Response) will play a role in the kind of Focus that an individual will construct.

In this ongoing work, the strength of the basic pattern remains pivotal and it has been a source of deep satisfaction to be involved in a chain of continuing exploration involving colleagues and graduate students, some of whose voices are represented in Edge and Wharton (2002, 2003). A more recent example is provided by Cheng (2006). On the topic of *'reading critically'*, a concept not necessarily as transparent as it might appear, she writes of how an SFRE awareness helps her develop such an ability. With regard to one methods article that she has read, her concluding comments are:

> The article is quite well-structured and seems to provide teachers with effective techniques to encourage students to talk with others. The problem I find is where the criteria or evidence are. When it comes to Evaluation, criteria or evidence are the main elements which the worth of suggestions can be judged by, or demonstrated with. However, the author concludes the article by just repeating her own thoughts without revealing either criteria or evidence to consolidate her arguments. Being conscious that the part of Evaluation is weak, I become doubtful about the techniques she tenders. To sum up, all these principles allow me to analyze and to criticize at the same time.
>
> (Cheng 2006: 12)

With regard to her own writing, she offers the following:

> So far I could recall how frustrated I felt when handing in the first draft to a professor. He just took a look at all the titles and subtitles on the draft and then told me to re-write it as there was no aim or purpose demonstrated in this writing. I still felt vague about how to organize the article and what should be included into it and what should not. The awareness of the SPRE or SFRE pattern helps me [with] how to formulate my ideas and in turn how to write more purposefully and coherently every time when I commence with assignments.
>
> (Cheng 2006: 13)

My own way forward with regard to further investigation of the *central dynamic* and its role in understanding and supporting course participants is yet to be walked. I am interested in seeing how useful a more detailed sub-categorisation of the central dynamic can be to teachers as they work to establish a Focus for the purposes of action research investigations (usually a difficult stage), and then the extent to which such increased specificity has an effect on their subsequent projects and their reporting of them.

Will this interest win through to be prioritised in my job? I do not know, but if I can pass on the research interest itself, and someone identifies it as an affordance for themselves, perhaps that will be a reasonable outcome of this reflection. In the meantime, I take great satisfaction from the mutually supportive 'passing on' already indicated above with regard to teaching and collegial collaboration.

At the same time, I have evoked the spirit of action research and that is my next topic. When referring to Hoey's analysis of semantic patterning, I drew on the use of *discourse analysis* as a term for dealing with the explicit data of linguistic communication. When I shift to action research, I mean to use 'discourse' rather in the sense of a 'world of discourse'. To use Gee's terms, I shall shift from the 'little d discourse' of language-in-use to the 'big D Discourse' of 'ways of being in the world':

> ...ways of acting, interacting, feeling, believing, valuing, together with other people and various sorts of characteristic objects, symbols, tools, and technologies – to recognise yourself and others as meaning and meaningful in certain ways.
>
> (2005: 7)

Action research

From my description, above, of my early use of Hoey's approach to semantic patterning in the interests of teacher education, it might just be possible to discern the outlines of action research, in spirit if not in formalised detail. This is because, while my typically pragmatic anglo-saxon attitude to theory and practice disposed me to seek knowledge in a problem-solving framework, I had not actually heard of action research in 1984.

I became more familiar with it as I pursued my studies and it has remained a source of inspiration and encouragement to me over the years, not only for its pragmatic strengths, but also for its lifestyle aspirations:

> participation in educational action research practices are particular ways of living and understanding that require more of the researcher than the 'application' of research methods.
>
> (Carson and Sumara 1997: xiii)

as well as the epistemological underpinnings embraced by some of its practitioners:

> suggesting that the locus of cognition and interpretation is not outside the individual human subject waiting to be interpreted (as psychological 'cognitivist' theories suggest) or embedded inside the individual (as 'constructivist' theories proclaim) but, rather, exists in the ever-evolving, complex joint actions among persons and their environments.
>
> (Carson and Sumara 1997: xix)

Methodologically, action research is conventionally represented as a continuing, interactive cycle of *planning, action, observation* and *reflection* (Kemmis and McTaggart 1988: 11–14). No one pretends that these elements are temporally separate stages of the work in any hermetic sense, but one can see a meaningful progression through them while also acknowledging that each permeates the others. Let me now elaborate a slightly more detailed version of my own current 'take' on action research, one that has clearly been moulded by a weight of experience of individual investigations into situations in which the researcher is also an engaged participant.

The whole significance of action research is that it begins in the context of one's ongoing professional *action*.

Here, one works with an aware openness to the work environment and with an attitude of wanting to learn from it, and to respond to it, with an overall purpose of making things better for all concerned. One accustoms oneself to being *observant* of what is going on and to be on the look-out for something worthy of further investigation. Stereotypically, this might be a problem, but it might also be a significant success, or something of interest for some other reason. The fundamental issue is that something significant is noticed by the individual in that person's working context. To put this another way, the context affords the person a particular opportunity, one that it might not have afforded another person, because that other person might perceive things differently.

The person concerned *reflects* on this issue and informs themself about it, drawing on personal experience, received knowledge, conversations with colleagues and also their own ability to articulate and to analyse the issue concerned.

Then comes a *planning* stage, personal but informed by others, by local knowledge and by knowledge drawn from the literature. Planning is also furthered by the articulation of emergent plans of investigative action as they are formulated.

On implementing the plan in *action*, one *observes* carefully what has changed and informs oneself as to whether things have improved or not from the various perspectives of the people involved, and whether one understands more. The articulation of such understanding as emergent theory is also a part of the action that comprises the ever-changing professional context, as well as potentially contributing to the literature of the field.

This thinking leads me to the following working definition of action research in my own practice:

> A rigorous investigation which sets out to improve the quality of experience and outcome available to participants in a given situation, while also enhancing their ability to articulate an understanding of what they have learned, and thus their potential to continue to develop in this and other situations, as well as their potential to contribute to the creation of knowledge.

There are various strands to this definition, each of them woven in in order to engage a different part of various ongoing arguments about, for example, the relationship between theory and practice, the relationship between the specific and the general, and the nature of knowledge and how one can contribute to it. The issue on which I want to focus on this occasion, however, is the relationship between this (I believe uncontroversial, although certainly not all-encompassing) definition of action research and the previous section on semantic patterning and discourse organisation. It has seemed to me for some time (Edge and Wharton 2003) that the SFRE pattern represents in the domain of discourse analysis the same essential framework for development that action research represents in the domain of research paradigms. In this sense, we can see a parallel dynamic functioning in both areas. (See Wharton, in press, for further discussion of this phenomenon.)

The similarity extends to the inhibiting fixation, in passing references to action research (see, e.g., Richards and Farrell 2005: 171) on problem-solving, even to the extent that researchers who wish to orient their work towards positive growth and development (e.g., Ludema, Cooperrider and Barrett 2001: 189) feel obliged to differentiate their work from action research *'because of its romance with critique at the expense of appreciation'*.

Understood in the light of an interplay between Situation and Evaluation, however, the central dynamic of an action research investigation can be triggered by whatever affordance is established between participants and their environment. We are back to the scenario sketched above of you, me, Tom, Dick and Harriett.

This is not to say that a one-to-one match is available between the pattern of SFRE and the cycle of Observation, Reflection, Planning, Action, but it is not a complicated job to zip them together: Acting in a Situation, one Observes a Focus on which one Reflects and to which one Plans a Response which one puts into Action and Evaluates.

With regard to what one passes on in educational encounters, the attitudes of action research and the discoursal capacity to express them can be very satisfying to see taken up. One such attitudinal statement from someone just beginning to come to grips with an action research approach is explicit and resonant:

> My notion of theory, which I suppose I had been carrying around with me since my teaching diploma, defined it more as an externalized body of knowledge to be imposed upon practice, so it is profoundly empowering to realize that pedagogic theorizing actually arises out of reflective practice within one's own local context. The value of this insight cannot be overstated, for the rejection of universal, totalizing theories seemed to connect with so many areas of the literature that I subsequently read.
>
> (Packett 1998)

So, if discoursal encounters mediated by Hoey's approach to semantic organisation have proved explanatory and powerful in the discourse community of action research, and the discourse of action research has proved inspirational in terms of personal and professional empowerment, how do I relate Cooperative Development to this closely interrelated duo? Before I attempt that, a brief introduction of Cooperative Development itself is called for.

Cooperative Development

Cooperative Development is one strand of expression inside TESOL teacher education of the idea that, in addition to approaches based on the ideas of copying external models and of applying received knowledge, we also need to encourage an internal growth approach to teacher preparation and continuing professional development.

Cooperative Development is a discourse framework in a sense that requires another slight shift in just what we mean by the term, discourse. Based on Rogerian principles of non-judgemental communication (Rogers 1969, 1980), Cooperative Development lays down rules of interaction for what its participants (a Speaker and an Understander) must and may not do. In this brief description, I capitalise terms that have a specific usage inside the Cooperative Development framework.

Cooperative Development requires Speakers to articulate their ideas and thoughts on a topic of interest for themselves in an attempt to move through stages of exploration and discovery to reach an insight on which they can base future action. They may not ask for, and must not expect, any advice, opinions or information from the Understander. Cooperative Development requires Understanders to put aside any ideas, experience or opinions of their own and to concentrate on Understanding the Speaker as fully as they can, sincerely making every effort to accept what the Speaker says and to empathise with the Speakers' viewpoint, while at the same time neither agreeing nor disagreeing with what the Speaker has to say. These two requirements create unusual pressures on both participants.

It is the claim of Cooperative Development that these pressures can help Speakers move their thinking forward in creative ways. The creativity is achieved because, in articulating one's thoughts in an open, non-defensive way, one does not only speak ideas that have already been fully formed, one also develops the ideas in and through Speaking. The Understander employs learned skills of non-judgemental discourse, including, centrally, the ability to Reflect back to the Speaker an accurate and sensitive version of what the Understander has heard, as a way of facilitating movement from the generality of talk to the specificity of action.

While Cooperative Development was not explicitly designed to fit the action research cycle, it is not difficult to see how it fits into the epistemological schema of experience-based learning and the ideological schema of self-empowerment that also inform action research. It also fits in, so to speak, methodologically at each stage of the action research cycle where articulation is mentioned in the above description (p. 238).

In terms of passing things on, and getting things back, this work in Cooperative Development reverberates for me in such texts as Boon (2003, 2005); Boshell (2002); Butorac (2006, under review); Mann (2002a, 2002b); de Sonneville (2007); Stewart (2003); and in its ongoing developments and uses, such as the email Cooperative Development session with Nur Kurtoglu-Hooton that helped me shape this chapter.

At this point, it seems to me that I have zipped together the action research cycle and the SFRE pattern, and also fitted Cooperative Development into the action research cycle. Where does that leave SFRE and Cooperative Development?

SFRE and Cooperative Development are both concerned with forms of discourse organisation and patterning, the former with regard to informing us about how discourse actually is organised in a certain community of practice, the latter with regard to extending the discourse repertoire with a view to serving that community further. If SFRE is analytic/descriptive, Cooperative Development is synthetic/interventionist; the former offers a key to valid peripheral participation (Lave and Wenger 1991) in the community, the latter offers a key to effecting change in the community. Together, perhaps, they counterbalance each other in the continuing effort to support stability and dynamism, and to celebrate unity and diversity (Edge and Wharton 2002) in the discourse community.

The good, the true and the ugly

In the previous three sections of this chapter, I have outlined three different types of discourse that I believe have been particularly significant in my professional life so far. They are areas in which I have tried, and am trying, to pass something on, for which effort I have received, and continue to receive, a great deal in return. I have also indicated what I see as connections and interactions between these three areas.

In this final section, I want to look at them again in a more all-embracing way, and to return to the issue of coherence.

I believe that the key to any coherence that my work might have turns around action research. Without wishing to sound too overblown, I believe that it connects to what I understand to be at least a part of the meaning behind Reason and Bradbury's (2001) evocative title: *Inquiry and Participation in Search of a World Worthy of Human Aspiration.*

This recognition in itself releases another storyline. The motivation that I know lay consciously behind all my early work as a teacher was to make a difference for that band of learners where what I could do might mean the difference between their passing and failing, between getting the place or not getting the place, being awarded the certificate or not, the learners who Appadurai (2006: 168) refers to as '*the bottom portion of the upper half*'. My inclination toward pragmatic exploration led me into a mode of work that I was to come to recognise as describable as action research. It seemed like a way to do good. If one can bear the simplicity of the distinction, I see action research as engaged less in a search for what is true, than for what is good. Let us pause on that one for a moment. I acknowledge the logic of the approach that says that we cannot begin to make something better until we know how it is now. We must, according to this argument, first attempt some approximation to the truth before we set about our improvements. There is, therefore, a necessarily sequential relationship involved between the true and the good.

However purely rational in the abstract (Clarke and Edge 2007) that logic might be, equally reasonable in situated terms is the response that says, 'Let's try to make things better here and now and see what we learn from that.' In arguing this, we come very close again to re-juxtaposing the two building blocks of Situation and Evaluation, which do not necessarily occur strictly in that order and which are not altogether separate from each other. After struggling with this thought for some time, I was delighted to find it so forcefully stated in the key publication already referred to:

> the primary purpose of action research is not to produce academic theories based on action; nor is it to produce theories about action; nor is it to produce theoretical or empirical knowledge that can be applied in action; it is to liberate the human body, mind and spirit in the search for a better, freer world.
>
> (Reason and Bradbury 2001: 2)

The claim that action research supplies the key to any coherence that my work has is supported by my having aligned myself in that area with Gee's (2005: 7) 'big D' Discourse as '*a way of being in the world*'. I now see that Cooperative Development and SFRE are two ways of languaging that world. Cooperative Development works through time and mediates between the energy of thought and the matter of language in the cycles of action. SFRE exists in space and provides one way of patterning the energy of action and thought in the linguistic matter of its representation.

However, I am left thinking, as action research seeks intersubjective confirmation that its investigations have led to improvement, and as

SFRE pushes cyclically on to a final representation of positive Evaluation, one issue that stands out is the non-judgemental nature of Cooperative Development. If Evaluation is the basic element of interaction with Situation that triggers the central dynamic of intervention, how does Cooperative Development function without it?

To recap: in SFRE, it is Evaluation that destabilizes Situation and precipitates the central dynamic of the text. Evaluation determines the nature of the trigger element of the dynamic and, therefore, both what will count as an appropriate response and what will count as appropriate criteria for the next Evaluation. The writer seeks to convince the reader of the efficacy of the dynamic that is triggered. The use of this language of persuasion, we note, although superficially monologic, is actually fundamentally interactive.

In Cooperative Development, it is the Understander's withdrawal of interactive judgement that turns the need for Evaluation back to the Speaker. Instead of supplying an other-evaluating dynamic of dialogic displacement, the Understander makes possible a self-evaluative, monologic dynamic of individual augmentation. Cooperative Development is, therefore, not lacking in Evaluation. The non-judgemental attitude of the Understander intensifies the Speaker's need for self-evaluation.

The main realisation that has arisen for me during the writing of this chapter has been this recognition of the central role of Evaluation in the discourse encounters that have become central to my work: how it determines the nature of the central dynamic in SFRE patterns, how it determines the nature of the affordance that precipitates the action research cycle, and how the non-judgemental attitude of the Understander in Cooperative Development redoubles the weight of Evaluation that falls on the Speaker so that, once again, it is the capacity of Evaluation to demand a response that powers development. Once made, furthermore, the distinction between Evaluation and being judgemental is clear to me. Evaluation is inescapable; how judgemental we are lies in our own hands. It is obvious. Discoveries are frequently like that.

Having noted the above points regarding the true and the good, it behoves me not to overlook the ugly.

One ugly truth that emerges is the unavoidable acknowledgement that my work, in terms of what I have passed on so far, has missed its mark. While the intention was to make a critical difference for those who might otherwise not get over their own particular bar, I recognise that my work has actually always been most useful to the most able. Perhaps this is inevitable, because I think, in retrospect, that I work most happily in a space of pragmatically oriented awareness-raising that is less easily accessible than I would have wished to those that I should most like to have helped. I cannot, however, find any point in regretting that now and note that I should, at this stage, probably concentrate on making the contribution that I am best suited to make.

To employ a framework famously constructed by Hymes (1972) and more recently evoked by Widdowson (2003), I believe that I have, in the context of TESOL action research, helped to develop two discourse tools (SFRE and Cooperative Development), the productive uses of which are possible, and for some people also feasible. Issues of personal preference and cultural context play into how appropriate they are, and it is an empirical question to what extent they actually happen.

Another ugly possibility, arising in the area of appropriateness, is the extent to which my comments above, especially about the SFRE pattern, can be seen as hegemonic in cultural-political terms. This is difficult. They sound hegemonic to me. On the other hand, I am articulating an understanding that makes sense of what I have experienced and what I have read. With regard to learning to function in the world of currently dominant academic discourse, I have yet to meet a student whose task was not eased at least somewhat by an awareness of the functioning of the SFRE pattern. But, of course, learning to fit in with the dominant discourses is a sure sign of the influence of hegemonic power. How to proceed?

Well, the other important parts of the deal, when extending one's repertoire to include mastery of dominant discourses, are first, to be aware that that is what one is doing; second, to do so without losing respect for the discourses that have shaped one's previous development, and third, to use that discoursal awareness to serve the good as one best understands it, including sometimes working to subvert generic norms as that seems appropriate.

Conclusion

I set out in this chapter to bring together what I see as significant styles of discourse encountered in a working life thus far in order to see if, from the perspective of now, any useful insights emerge. I find that they have, and that they help me construct a view of what it is that I would like to pass on through the work that I continue to do as teacher, supervisor, writer and colleague:

- a commitment to action research as a world of discourse in which the overriding commitment is to the *good*, accompanied by a continuing exploration of the relationship between the good and the true;

- a desire to understand better the interactions between, on the one hand, the fundamental elements of Situation and Evaluation and, on the other, the various manifestations of the central dynamic that are set in motion by them as we use the SFRE pattern to schematise our constructions of the action research that we carry out, and in order to communicate our processes and our discoveries;
- a desire to understand better the workings of non-judgemental discourse in Cooperative Development, and to employ that understanding more effectively at the service of personal and professional development, individual and collegial, as we take our projects forward in interaction with the communities that we seek to build.

These are not small goals and I am all too aware how poorly I sometimes embody them. I can only hope and believe that one's failings do not invalidate one's aspirations. I have, anyway, found it personally useful at this stage of my working life to reflect on whether there is anything else to say about the coherence of my discourse experiences in TESOL above and beyond Gee's observation that:

All life for all of us is just a patchwork of thoughts, words, objects, events, actions and interactions in Discourses.

(2005: 7)

I thank the editors for this opportunity and I hope that others may construct something useful for themselves out of their encounter with this text.

Acknowledgements

In addition to the editors, I wish to thank Susan Brown, Richard Fay, Nur Kurtoglu Hooton, Michael Hoey, Gary Motteram, Diane Slaouti, Juup Stelma and Sue Wharton for their comments at various stages of the production of the above text.

References

Appadurai, A. 2006. 'The right to research'. *Globalisation, Societies and Education*, 4(2): 167–77.
Boon, A. 2003. 'On the road to teacher development: Awareness, discovery and action'. *The Language Teacher*, 27(12): 3–7.
Boon, A. 2005. 'Is there anybody out there?' *Essential Teacher*, 2(2): 1–9.

Borg, S. 2006. *Teacher Cognition and Language Education: Research and Practice.* London: Continuum.

Boshell, M. 2002. 'What I learned from giving quiet children space'. In Johnson and Golombek, pp. 180–94.

Butorac, D. 2006. 'Learning through talk: An evaluation of cooperative teacher development'. Unpublished MA dissertation. Sydney: Macquarie University.

Butorac, D. (under review) 'Curriculum change as teacher development'. *Prospect.*

Carson, T. and Sumara, D. 1997. 'Reconceptualizing action research as a living practice'. In T. Carson and D. Sumara (eds), *Action Research as a Living Practice.* New York: Peter Lang Publishing, pp. xiii–xxxv.

Cheng, J.-M. 2006. 'Awareness-raising through analyzing written discourse'. Unpublished independent study paper, MA in Ed. Tech. and TESOL, University of Manchester, England.

Clarke, M. and Edge, J. 2007. 'Seeking reason in rational and reactive times'. Paper presented at CA-TESOL Annual Convention, San Diego, April 2007.

de Sonneville, J. 2007. '"Acknowledgement" as a key in teacher learning'. *ELT Journal* 61(1): 55–62.

Edge, J. 1989. 'Ablocutionary value: On the application of language teaching to linguistics'. *Applied Linguistics* 10(4): 407–17.

Edge, J. 1992. 'Cooperative development'. *ELT Journal* 46(1): 62–70.

Edge, J. 1994. 'Empowerment: Principles and procedures in teacher education'. In R. Budd, D. Arnsdorf, and P. Chaix (eds), *Triangle XII: The European Dimension in Pre- and In-Service Language Teacher Development – New Directions.* Paris: Didier Erudition, pp. 113–34.

Edge, J. 2001. 'Attitude and access: Building a new teaching/learning community in TESOL'. In J. Edge (ed.), *Action Research: Case Studies in TESOL Practice.* Alexandria, VA: TESOL, pp. 1–11.

Edge, J. 2002. *Continuing Cooperative Development: A Discourse Framework for Individuals as Colleagues.* Ann Arbor: Michigan University Press.

Edge, J. 2006. 'Non-judgemental discourse: Role and relevance'. In J. Edge (ed.), *(Re)Locating TESOL in an Age of Empire.* Basingstoke: Palgrave Macmillan, pp. 104–18.

Edge, J. and Wharton, S. 2002. 'Genre teaching: The struggle for diversity in unity'. In K. Miller and P. Thompson (eds), *Unity and Diversity in Language Use.* London: BAAL/Continuum, pp. 22–38.

Edge, J. and Wharton, S. 2003. 'Research in teacher education: Reading it, doing it, writing it'. In B. Beaven and S. Borg (eds), *The Role of Research in Teacher Education.* Whitstable: IATEFL/Oyster Press, pp. 49–53.

Edge, J. and Richards, K. (eds). 1993. *Teachers Develop Teachers Research.* Oxford: Heinemann International.

Gee, J. 2005. *An Introduction to Discourse Analysis,* 2nd edn. London: Routledge.

Hoey, M. 2001. *Textual Interaction: An Introduction to Written Discourse Analysis.* London: Routledge.

Hoey, M. 1983. *On the Surface of Discourse.* London: George Allen & Unwin.

Hymes, D. 1972. 'On communicative competence'. In J. Pride and J. Holmes (eds), *Sociolinguistics.* Harmondsworth: Penguin, pp. 269–93.

Johnson, K. and Golombek, P. (eds). 2002. *Teachers' Narrative Inquiry as Professional Development.* Cambridge: Cambridge University Press.

Kemmis, S. and McTaggart, R. (eds), 1988. *The Action Research Planner*, 3rd edn. Geelong: Deakin University Press.

Lave, J. and Wenger, E. 1991. *Situated Learning: Legitimate Peripheral Participation*. Cambridge: Cambridge University Press.

Ludema, J., Cooperrider, D. and Barrett, F. 2001. 'Appreciative inquiry: The power of the unconditional positive question'. In Reason and Bradbury, pp. 189–208.

Mann, S. 2002a. 'Developing discourse in a discourse of development'. Unpublished PhD thesis, Aston University, Birmingham.

Mann, S. 2002b. 'Talking ourselves into understanding'. In Johnson and Golombek, pp. 195–209.

Packett, A. 1998. Unpublished response to Foundation Module task, MSc in TESOL, Aston University, Birmingham, England.

Reason, P. and Bradbury, H. 2001. 'Inquiry and participation in search of a world worthy of human aspiration'. In Reason and Bradbury, pp. 1–14.

Reason P. and Bradbury, H. (eds). 2001. *Handbook of Action Research*. London: SAGE.

Richards, J. and Farrell, T. 2005. *Professional Development for Language Teachers*. Cambridge: Cambridge University Press.

Rogers, C. 1969. *Freedom to Learn*. Columbus, OH: Merrill.

Rogers, C. 1980. *A Way of Being*. Boston, MA: Houghton Mifflin.

Stewart, T. 2003. 'Insights into the interplay of learner autonomy and teacher development'. In A. Barfield and M. Nix (eds), *Autonomy You Ask!* Tokyo: JALT Learner Development SIG, pp. 41–52.

Van Lier, L. 1996. *Interaction in the Language Curriculum*. Harlow: Longman.

Wharton, S. 1999. 'From postgraduate student to published writer: Discourse variation and development in TESOL'. Unpublished PhD thesis. Aston University, Birmingham.

Wharton, S. 2001. 'Writing from a context: Course assignments and professional development'. Paper presented at *Writing Development in Higher Education*, University of Leicester, 24–25 April 2001.

Wharton, S. (in press). 'Social identity and parallel text dynamics in the reporting of educational action research'. *ESP Journal*. doi:10.1016/j.esp.2006.09.003

Widdowson, H. 2003. *Defining Issues in English Language Teaching*. Oxford: Oxford University Press.

Winter, E. 1986. 'Clause relations as information structure: Two basic text structures in English'. In M. Coulthard (ed.), *Talking about Text*. Birmingham University: English Language Research, pp. 88–108.

Zembylas, M. 2005. 'Beyond teacher cognition and teacher beliefs: The value of the ethnography of emotions in teaching'. *International Journal of Qualitative Studies in Education*, 18(4): 465–87.

Reflections on Passing
on the Knowledge

Fotini Vassiliki Kuloheri

Introduction

'Passing on the Knowledge' implies, from my perspective, being actively involved in imparting useful knowledge in effective ways that can make the value of the knowledge evident and the learning experience beneficial to the recipients. As an ELT practitioner lucky to have reached this stage and prompted by Kuchah's, Edge's and Wharton's chapters to recall career memories of a lifetime, I have realised in retrospect that events, situations and actions have been so closely interwoven that it is unavoidable, while focusing on this main stage, not to also recall the former stages, the discourses of which have shaped my current professional activities substantially. From the three preceding chapters, it is mainly Kuchah's that guided my recollections of all the stages above due to its detailed autobiographical nature. The other two have stimulated experiences related to the particular topics they pose, that is, semantic patterning, Cooperative Development, Action Research and writing for publication.

My personal 20-year TEFL history has taken place in Greece, where a prominent role is attributed to EFL for the same reasons that apply to the Cameroonian context, that is, the leading social, economic, technological and political role of English globally (for a thorough discussion of the background to the preference of EFL in Greece, see Prodromou 1990). My career development can be seen as a continuous shift from the practical to the theoretical and vice versa during efforts to establish my own niche in the TEFL areas I have been exploring.

Similar to Kuchah, my passing-on-knowledge approach has been determined by a series of contexts, each one of which had its own integrity, but still they have all managed to shape my current beliefs, decisions and actions. Most of all, it has been two training courses I attended, that is, one by the Institute for Applied Language Studies of the University of Edinburgh leading to a Certificate in Advanced Studies in ELT and another one by the University of Cambridge LESIE leading to the Royal Society of Arts Diploma for Overseas Teachers of English.

The pros and cons of these courses have defined my current teacher training practice as follows:

1. Training on the main course focus is combined with a thorough practical revision of the aims and associated activities of the teaching of vocabulary, grammar and communicative skills.
2. Theory is combined with practice, methodology with language development and input in sessions with assignments, teaching practices, and peer-observations.
3. Time constraints and syllabus requirements are usually negotiated so that they do not keep a tight rein on teachers' practices and allow for reflection.

Kuchah's reference to educational policies not tailored to local realities and governed by political ambitions have given rise to my own relative experiences, which were acquired during the second cycle of my career from my roles as an EFL primary state school teacher, an educator shifted to an administrative ministerial position (a very frequent prospect for Greek state educators) and a doctoral student.

But in order to articulate how these experiences have affected my professional development, I believe it is necessary to establish a common framework of reference by presenting briefly the challenging discursive contexts in which they occurred.

Limitations of the educational system

My appointment to a permanent position as a state school EFL teacher came as a relief in terms of the security of the employment (in contrast to my constant dismissal and re-employment in the private sector at the end and the beginning of each school year respectively) and of the relaxed teaching pace the educational system permits (in contrast to the restlessness caused in the private sector by the demand for increased qualifications). Nevertheless, my restless nature started picking out disadvantages deeply seated in the educational system, as reflected in the following indicative questions.

How can pupils consolidate, memorise and expand EL knowledge when lessons take place only three times a week? How can we face the usually low educational attainment level of the increasing number of immigrant children attending Greek state schools? What is the cultural background of these immigrant pupils, and how can we inform our teaching practices to cater for the respective learner needs? In what ways are the Greek prefectures differentiated culturally and what are the implications for our teaching? How can we satisfy the exam-oriented learning targets of pupils and parents, when the curriculum and the course books do not promote the achievement of this aim? How can we gain the attention and interest of the majority of pupils who attend EFL exam-oriented evening classes at private foreign language institutes (Mattheoudakis and Nicolaides 2005)? What teaching techniques can we use to implement the newly introduced cross-curricular approach to TEFL (Government Gazette 2003)? What discipline strategies can we employ to restrict classroom indiscipline and the consequent teacher dissatisfaction?

Bureaucracy and political aspirations

Even more than the practical teaching experience, the administrative position in a state educational institute I was shifted to (a common and much favoured prospect for Greek educators) gave me a golden opportunity to gain a much deeper insight into the individuals and the processes that determine educational change.

A first challenge was the difficulty of responding to the demands of the administrative tasks without prior training, to which I rose through the consultation of experienced administrative employees and colleagues formerly shifted to the position and through the adoption of an exploratory approach comprising continuous deductive and inductive mental processing.

The particular challenge helped me evaluate as negative the Ministry's policy to permit that Greek experienced and qualified teaching staff, who should be used in didactic practice, remain unexploited in posts of a purely bureaucratic nature. But more important than that, I became aware of Kuchah's realisation that 'bureaucratic complexities tend to widen the gap between pedagogic authorities and the classroom procedures'.

The second challenge derived from discursive interactions during my service to various educational leaderships. From this experience I have found that the personality of each leader plays a crucial role in policy-making. Specifically, individuals exerting centralised and absolute authority hinder the development of the skills of their subordinates and, consequently, decrease pluralism and obstruct innovations, in contrast to communicative and cooperative individuals who tend to decentralise authority by distributing responsibilities.

Also, I felt demotivated when I understood that a limited number of capable, ethical and enlightened individuals at the top of an educational system is not enough to lead change; what counts more is the synthesis of the whole group of educational leaders, which should instil mutual trust and an agreement on the targets set and the processes to be followed. Eventually, like Kuchah, I was left with the frustrating awareness that educational policy-making is guided mainly by political ambition and/or by a general and imprecise perception of Greek school reality.

Consequent new teacher training perspectives

As a consequence of the above limitations – coupled with the fact that state school EFL advisors are very few and, therefore, not capable of providing adequate and effective guidance – and of the recent doctoral studies I have undertaken in Education, I have come to the conclusion that teachers are basically left alone to face the complexities of their daily tasks, and that what can empower them to do so can be a coherent, realistic training plan in the form of pre-service and in-service courses carried out in local school contexts by trainers with knowledge about, and experience in, state education, TEFL methodology and educational research.

Specifically, extending the implication of the Situation-Problem/ Focus-Response-Evaluation for the remedy of teaching/learning dysfunctions, I believe that the pattern can inform a training approach aiming at context-specific solutions to EFL problems. From this perspective, teacher education can include training in research traditions and methods that facilitate the close study of a Problem in a specific classroom/school Situation with the purpose of reaching a Response that will ensure a positive Evaluation of the improved Situation.

Edge has discussed the advantages of AR, which he is fully justified to have referred to extensively. AR is, in my experience, not utterly new in the sense that it has been present in the form of *active teaching practice* in the cases of teachers who faced their profession with a strong sense of responsibility and enthusiasm and, as a result, adopted a continuum of teaching, evaluating, re-planning and re-trying. What is new is, in my view, the systematised nature it acquired as a research method of inquiry, the inclusion of which in the syllabus of teacher development courses can renew the respective initial intuitive process, establish it on a different, methodical basis, and thus invite an increased sense of responsibility and seriousness towards the profession.

In addition to AR, case study (CS) can be considered another useful research tradition for teachers who want to examine systematically, interpret and understand deeply the interaction of important factors central to a specific routine and problematic area of interest in their professional territory (e.g. Bassey 1999; Stake 1995) without separating its variables from their context. Expanding on Edge's line of thought, since there is admittedly a necessary succession of the true by the good, and since in my opinion CS seeks the deep truth, whereas AR seeks the useful, the two traditions can be regarded, from my perspective, as supplementary too.

Finally, some basic features of CD expounded on by Edge, as I experienced them in the professionalism of my postgraduate tutors, have reinforced in my training approach the encouragement of trainees to act as Speakers and my performance as an Understander.

Designing TEFL material

In 1995, the broadening of my horizons after the completion of my MA studies in Applied Linguistics in Essex University gave rise to my preference in teaching EFL to young learners and, specifically, in focusing on the use of audiovisual material for the children's language development (e.g., TV programmes and videos/DVDs), the teaching of grammar and vocabulary at primary ages and the error types in the English spelling of young Greek EFL learners.

The focus of my dissertation, that is, the evaluation of existing EFL video materials for this age group and the development of a series of linguistic criteria for their design, resulted in my strong wish to contribute to the design of printed EFL material too. This wish was fulfilled when my cooperation with Oxford University Press was launched for the authoring of the first improved version of a type of self-access material supplementing a course book series of the publishing house used in Greece.

The knowledge I have been passing on during the publication of eight books so far (see, e.g., Kuloheri 1994, 2005) relates to what children need to become more effective and autonomous language learners within the context of each respective series used, as these needs are expressed by the children themselves, their teachers and their parents. So, besides seeing this new area of development like Kuchah as a creative chance to invest my accumulated TEFL knowledge and classroom experience, for the purpose of contributing to the achievement of the relevant learning goals, I also defined its role from the perspective of the deficiency view (Allwright 1990), that is, to remedy the deficiencies of the teaching process, like insufficient explanation of vocabulary and grammar and inadequate language practice.

My involvement in material design was expanded to other types of EFL material when I undertook to write a grammar book reinforcing the Presentation-Practice-Production approach in the teaching of grammar (Kuloheri 2003). The result has been a book that, according to colleagues, succeeds in enhancing an effective learning process comprising the steps of first understanding new concepts and functions, then recognizing meanings and uses, and then using them actively for communication purposes by means of controlled and less controlled tasks.

Writing for publication

Writing for publication has been a desire of mine ever since I completed my MA studies, and has been intensified during my doctoral studies. This is, in my opinion, due to the fact that these studies helped me develop the expertise Wharton mentions at the end of the section 'Some challenges in becoming a writer', familiarised me with the discourses of theories and research and with their relevance to my accumulated teaching experience, and raised awareness in me of my capabilities and interests.

Nevertheless, to reach the finished product and, in the long run, to become a published writer, I have been faced with difficult challenges, the major ones being limited time available due to commitments as a practising teacher, daughter of an old mother, wife and mother of two young children, the absence of encouragement and guidance in the production of the suitable text type, and my fear (confirmed by Egbert 2007, in Wharton's chapter) of the domination of certain kinds of writers in TESOL publications.

In regard to the first problem, I am experiencing the management of my life constraints as a necessary demanding continuous effort that takes me through various transitional stages. As for my problem with the absence of encouragement and guidance, this is obviously confirmed by the experiences evidenced by Wharton of teachers attending the distance learning MA course in Aston University in the form of the What? and How? questions posed. In the interest of space and given the fact that Wharton has adequately presented these problems, I intend to proceed beyond the problems to what need be done so that EFL teachers who have the necessary background knowledge and teaching experience will manage to develop the new identity of a writer successfully.

My suggestion, which I believe is essentially in agreement with Wharton's, is that a new module be offered in all BA, MA and doctoral schemes on professional writing in the sense of what Chouliaraki and Fairclough (1999; in Wharton) call 'a theoretical practice'. This module, in my opinion, should provide students with the background knowledge of discourse, semantics and pragmatics required through the close study and analysis of sample TEFL articles, as well as give them opportunities for the systematic guided development of at least one article for publication. Furthermore, it should aim at increasing their self-confidence and, through this, help them to silence the inner discouraging voices that in my experience raise obstacles during the writing process. For this purpose, it may be necessary that tutors have some background knowledge in psychology and sociology too.

I can already hear some saying: BA level is too soon to expect students to write for publication. Although this may well be so, especially because BA students may lack the teaching experience required, we should always consider the cases of experienced teachers who may never undertake postgraduate studies, but who may well wish at a certain stage of their career to submit a paper for publication. I would find it unfair for them to retain the single identity of practising teachers and be deprived of the better position that, according to Chouliaraki and Fairclough, expertise in professional writing can give them.

Conclusion

Kuchah's, Edge's and Wharton's chapters stimulated memories of thoughts, events, actions and interactions (Gee 2005, in Edge) in my TEFL career and helped me, thus, to become aware of the resulting pursuits at my current stage of 'Passing on Knowledge' as follows:

1. As a teacher trainer:
   • to link practice with theory and combine input on TEFL methodology and the EL system with teaching practice and reflection

- to adopt a trainee-centred approach that allows for the time and space required for the investment of the trainees' background knowledge and experience during reflection and decision-making
- to provide solutions tailored-made to the contexts in which teaching/ learning problems occur through the implementation of the SPRE/SPFE pattern
- to become more effective in my extended role as an ELT consultant
- to contribute to the development of the role of the pedagogical researcher in EFL teachers through my participation in teacher training programmes focusing on Action Research and Case Study, with a very strong practical component which will give trainees the chance to put their hands on various traditions and methods, observe others conducting research and receive guidance and feedback in their own investigations from experienced researchers in the field
- to implement and explore the Speaker–Understander relation of Cooperative Development during training sessions.
2. As an EFL materials designer to continue serving learner needs for the purpose of facilitating the deep understanding of EL and the development of accuracy and fluency in language use.
3. As a prospective published writer to persist in my efforts to manage the constraints of my personal life, to explore the advantages of the Situation-Problem text pattern in writing and to take the risk of submitting my written work to refereed journals.

Certainly, the above targets shaped by the discourses I have been exposed to indicate a long way ahead, requiring vigour, determination and commitment to the principles of quality teacher education captured by Alatis (2004: 31) in his motto 'Be professional. Get a good education. Provide service.'

References

Alatis, J. E. 2004. 'The psychic rewards of teaching'. *English Today*, 20: 30–7.

Allwright, R. L. 1990. 'What do we want teaching materials for?' In R. Rossner and R. Bolitho (eds), *Currents of Change in English Language Teaching*. Oxford: Oxford University Press, pp. 131–47.

Bassey, M. 1999. *Case Study Research in Educational Settings*. Berkshire: Open University Press.

Government Gazette of the Hellenic Democracy. Issue B, Nr 303 and 304/ 13–03–03.

Kuloheri, F. 1994. *You and Me, Study Companion 1 and 2*, Oxford: Oxford University Press.

Kuloheri, F. 2003. *Grammar Team 1*, Oxford: Oxford University Press.

Kuloheri, F. 2005. *Rainbow, Study Companion 1 and 2*, Oxford: Oxford University Press.

Mattheoudakis, M. and Nicolaidis, K. 2005. 'Stirring the waters: University INSET in Greece', *European Journal of Teacher Education*, 28(1): 49–66.

Prodromou, L. 1990. 'English as cultural action'. In R. Rossner and R. Bolitho (eds), *Currents of Change in English Language Teaching*. Oxford: Oxford University Press, pp. 27–40.

Stake, R. E. 1995. *The Art of Case Study Research*. Thousand Oaks, CA: Sage.

Index